AF845201

CONVERGENOMICS

Convergenomics

Strategic Innovation in the Convergence Era

Sang M. Lee and
David L. Olson
University of Nebraska-Lincoln, USA

Routledge
Taylor & Francis Group
LONDON AND NEW YORK

First published 2010 by Gower Publishing

2 Park Square, Milton Park, Abingdon, Oxfordshire OX14 4RN
711 Third Avenue, New York, NY 10017

Routledge is an imprint of the Taylor & Francis Group, an informa business

First issued in paperback 2018

Copyright © Sang M. Lee and David L. Olson 2010

Sang M. Lee and David L. Olson have asserted their moral rights under the Copyright, Designs and Patents Act, 1988, to be identified as the authors of this work.

All rights reserved. No part of this book may be reprinted or reproduced or utilised in any form or by any electronic, mechanical, or other means, now known or hereafter invented, including photocopying and recording, or in any information storage or retrieval system, without permission in writing from the publishers.

Notice:
Product or corporate names may be trademarks or registered trademarks, and are used only for identification and explanation without intent to infringe.

Gower Applied Business Research
Our programme provides leaders, practitioners, scholars and researchers with thought provoking, cutting edge books that combine conceptual insights, interdisciplinary rigour and practical relevance in key areas of business and management.

British Library Cataloguing in Publication Data
Lee, Sang M.
Convergenomics : strategic innovation in the convergence era.
1. Organizational change. 2. Technological innovations--
Economic aspects. 3. Technological innovations--Social
aspects. 4. Social change.
I. Title II. Olson, David L.
658.4'062-dc22

ISBN: 978-0-566-08936-7 (hbk)
ISBN: 978-1-138-37813-1 (pbk)

Library of Congress Control Number: 2009941218

Contents

List of Figures

List of Tables

Foreword

Today, we live in a global community where people are connected and economies are interdependent. A small ripple in one corner of the world may cause a gigantic wave in many other places, as we have seen during the recent global economic crisis. Changes in the global economy are diverse, dynamic, and quite often unpredictable. To understand the root causes of these changes and predict how things will be, we must study historical developments, current events, and possible future consequences. This book, *Convergenomics: Strategic Innovation in the Convergence Era,* is a futuristic book that is written to shed light on the dynamics of the global economy. Two distinguished scholars, Sang M. Lee and David Olson, present a very readable book which is based on thorough research and real-world examples to make their points.

The modern world is based on a series of earth shaking developments, most notably the three waves – the agricultural revolution, the industrial revolution, and the information revolution. Lee and Olson present the Convergence Revolution as a fourth wave which is changing the way people live and behave, how organizations function and produce value, and how governments serve the people. Convergence is usually understood as the synergistic combination of different concepts or objects for other uses. *Convergenomics,* as the authors coined the word, represents a new approach or environment where new value can be created through a convergence process. Convergence is really nothing new. Throughout human history, people have learned to create new things through convergence. What is new is that the maturing infrastructure which now enables the magnitude and new possibilities of creating an explosion of new value through convergence. We have seen revolutionary developments in many isolated areas of human endeavors – information and communication technologies, biotechnology, nanotechnology, management innovations, globalization, and newly emerging economies, to name a few. However, today all of these major developments, which the authors call megatrends, have reached such stages that they now engender explosive possibilities for convergence. The authors give an ample number of examples of breakthrough inventions and innovations enabled by convergence in such diverse fields as IT, biotechnology, health care, new product and service development, education, entertainment, and so on.

In *Convergenomics*, Lee and Olson provide an insightful analysis of the most important megatrends that are shaping the stage for revolutionary convergences – globalization, digitization, changing demographics, changing industry mix, commoditization of processes, new global value chains, emerging economies, deteriorating environment, and cultural conflicts. The authors explain how the convergence revolution is shaping a new environment through organizational functions, interorganizational relationships, technologies, industries, open innovation, and bio-artificial systems, and they provide numerous examples of how human systems are transformed by *Convergenomics*.

The Gallup Organization is one of the world's largest collectors and users of data, information, and knowledge. Gallup has studied organizations over decades with the intent of identifying factors leading to organizational success. We converge the collected intelligence with unique work and living environments throughout the globe. We believe the ultimate convergence should be at the human to human level through convergence of emotional intelligence. Gallup's research and global consulting activities involving leadership development, employee engagement, well-being, cultural harmony, positive psychology and strengths-based management are all about human to human convergence. Professor Lee, who is a Senior Scientist of Gallup, and Professor Olson briefly discuss this most important yet complex aspect of convergence in this book. They have provided a focus on how organizations have been able to combine developments in technology and innovation to improve the quality of life for their consumers. I am certain that their continuous research will bring new insights and understanding about how true synergistic value can be created at the pinnacle of the convergence revolution. I do recommend this book to any leaders or aspiring leaders in business, non-profit, and governmental organizations.

James Clifton, Chairman and CEO
The Gallup Organization

1 Prelude to the Wave of Convergence

Trends

Today we live in a networked global economy. Behind the rapidly changing world, there are megatrends that are shaping the way that people behave and organizations work. Globalization and digitization are the two most obvious trends that have helped flatten the world as Thomas Friedman asserted. Globalization, through increased interdependence among nations, has greatly accelerated because of the advances in information and communication technologies (ICTs). Technologically advanced countries, such as the US, the UK, Germany, Japan, South Korea, and Finland, use IT tools to conduct global business (digitization). Even many underdeveloped nations take advantage of wireless mobile devices to connect to the global community. Anyone who has access to the internet can use the web to be in contact with anyone around the world, uncensored for the most part. Global e-business also allows people to transact with others through e-Bay or Amazon, for example.

Globalization has become a reality because of the networked global economy. Instead of a collection of closed economies, the world's economy has become transparently open, as the recent global financial crises attest. Through the enabler of digital technology, organizations can now span the globe. Operations such as those of Wal-Mart, IBM, and Nike draw upon contributors of world-class competencies from all over the world, and also sell to anyone with access to the internet. Newly developing technologies, such as nano- and bio-technology offer further dramatic new means to create value for customers. This globalization has provided a new global age where we share economic interests with everyone on the globe, helping overcome many of the ideological conflicts that have disrupted human progress in the past.

Many observers have noted the impact of digitization on progress. Moore's Law proposes that computing power doubles about every 18 months, with constant costs. Computer scientists have been phenomenally successful in providing us with ever more powerful computers. Closely related, Hwang's Law argues that flash memory chip densities will double every 12 months. Computers are organized into networks linked by the internet. Metcalfe's Law contends that the value of a network equals the square of the number of users. Essentially, the more people linked, the greater the value to us all. The impact of these developments on economic entities was addressed by Coase's Law, which states that as transaction costs decrease, firm complexity diminishes, and firms tend to expand until the cost of a marginal transaction internally equals the cost of the same transaction external to the firm. If the cost of working inside a firm gets too high, we have the ability to outsource.

There are other profound social, economic, and political undercurrents in play. For example, changing demographics now present a much different world map with new opportunities and looming social/political problems. The rapidly increasing older generation and the drastic decrease in the younger generation in most developed and developing countries present a new set of challenges. Simultaneously, in the Middle East, where unemployment is already rampant, population increase is outpacing what society can accommodate, thus creating explosive social problems for the future.

The changing industry mix in the developed and rapidly developing countries is another major force. The economic center of gravity has moved from agriculture through manufacturing to knowledge-intensive service industries. Japan has shown the world how to attain excellence in manufacturing. Asian Tigers, such as South Korea, Taiwan, and Singapore have kept the competitive pressure on Japan. In the United States and the European Union (EU), service industries have become dominant, especially in such high-value creating industries as healthcare, biotechnology, financial services, ICTs, recreation, hospitality, education, and entertainment. BRIC nations (Brazil, Russia, India, and China) are evolving into major players in the world's economy.

Related to the changing industry mix, there has been significant movement towards deregulation and privatization. In the United States, the Reagan presidency saw major initiatives in deregulation, exemplified by the airline industry and the communication industry. Instead of subsidizing archaic industries, the market leads us to abandon unproductive efforts and redirect energy to satisfy newly developing needs. Similar changes in the UK during the Thatcher era saw major movements towards privatization. The argument is that deregulation and privatization rely upon market forces to determine resource allocation, more efficiently than public policy makers can do. Thus, deregulation and privatization have become widely spread throughout the world. The current global financial crisis has raised the question as to whether or not deregulation in the financial industry has been overdone.

Another important megatrend is commoditization of processes. Through standardization and quality certification, operational processes become scalable. Standardization and clearly stated specifications enables the outsourcing of production. Thus, bundling, unbundling, and re-bundling of processes are feasible. Such approaches create new opportunities for global outsourcing and interorganizational collaboration.

Friedman[1] cited three convergences that enabled a flat world: (1) new players (through global access); (2) a new playing field (the web economy); and (3) the ability to develop new processes and ways of accomplishing work through horizontal collaboration. We agree with his assessment. In the new business environment, a single business organization often finds that it is impossible to secure all of the necessary core competencies to sustain a competitive advantage. Thus, developing new value networking through collaboration with partner organizations has become another global trend. Such collaborative inter-organizational relationships could be for process innovation, customer value creation, or new customer base development. Examples include healthcare network alliances, where each integrated patient unit (hospital) specializes in one or more types of treatment, referring other cases to alliance partners.[2] However, the world is not universally (or uniformly) flat. Rapid technological advances have widened the digital divide. Only 40 per cent of the people in Africa have internet access, and a typical rate is over $250 per month.

The global scene is also changing dramatically as the emerging new economies (for example, BRIC) grow. These countries are no longer playing supporting roles to the developed countries of the West. They are in fact leading the world in foreign exchange reserve (China), mid-tech industries, consumer goods, and even in ICTs, biotechnology, and ubiquitous computing. Of course continued economic emergence in south-east Asia plays a major role in the world economy, with the promise of increased contribution from developing countries.

There is a growing concern about the deteriorating environment, especially with regard to global warming and pollution. There is wide ranging public debate about the impact of global warming among political leaders, scientists, business organizations, and the public in general. While global warming and "green business" are beginning to receive serious attention from businesses, they will increasingly present both enormous challenges and opportunities as governments around the world join advocates of environmental well-being in enacting stringent rules dealing with this issue.

There is another explosive trend which has a major impact on the global economy. Increasing cultural conflicts are getting ever stronger while globalization and advanced ICTs tend to flatten the world. The standardization of communication media and global business practices has reinforced a strong, and sometimes violent, opposing force. For example, the internet is a force of democratization as it is color-, class-, gender-, and distance-blind. Thus, in those countries where traditional values or religions oppose equalization, globalization and ICTs are blamed for cultural contamination. Cultural conflicts often fan geopolitical conflicts that have brought regional wars, terrorism, and religious antagonism. The very forces that are making the world flat are also creating new barriers and deep canyons across cultures.

The megatrends described above have varying degrees of impact on the global economy. One of the most consistent trends has been the convergence of manufacturing and service. These two terms were used in the past to differentiate those who made things, focusing on cost minimization and efficient use of resources, to those who provided services to other people, focusing on customer satisfaction. This dichotomy seems irrelevant today, as no manufacturer of any product can focus solely on making its product efficiently without consideration of the service provided to the consumer. Boeing, for example, has to spend significant effort on providing better service and even steel manufacturers have to respond to the needs of those who purchase their products.

The most profound of the impacts of megatrends occurs through their interaction and interplay. Thus, *convergence revolution* is at the center of the whirlwind of global change. Convergence of business functions through business process re-engineering (BPR) widely practiced in the 1990s brought about process innovation. Convergence of technologies, organizations, industries, and people create new collaborations, open-source participation, new technologies, new products and even cybernetic organisms (cyborgs). People who are scattered around the world can share, create, and apply knowledge. Also, many seemingly heterogeneous technologies such as ICT, nanotechnology, biotechnology, neuroscience, and so on. can converge to create new products, services, or even technologies. Organizational convergence has become a popular strategy for many firms to leverage the advantage of open-sourcing, outsourcing, or global supply chain alliances. Industry strategic innovation involves the development of sustainable economies of convergence for competitive advantage.

The Adaptive Network of Ideas

The convergence trends have been observed in technology as well as in consumer product development. It is the meeting of demand (customer needs) and means (convergence). Human history is a continuous link of ideas and experiments building upon each other, connected in a chain of innovation in the light of new opportunities. One simple example is seen in telephony. Alexander Graham Bell developed a telephone system over 100 years ago. AT&T was granted a monopoly to develop this technology, initially linking the fortunate few who could afford access and who lived close to the network of lines AT&T created that linked telephones. Some of us remember when we had to crank telephones with numbers like "two longs and a short." Technology advanced to the point that dial phones were developed, with a coded numbering system covering the country, and then the world. This was followed by the walkie-talkie, initially a military tool to allow radio links over short distances between two transmitter/receivers. As the technology evolved, there was a convergence with the demand for consumer mobility, leading to the cell phone. Providers invested in cellular transmission networks. With the breakup of AT&T, many cell phone companies emerged, uniting wireless cell phones to the backbone telephone system. This evolution continues through convergence with internet technology in voice over internet protocol (VoIP), and tremendous expansion of the services available over cell phones, to include paging, texting, and internet access allowing e-business, as younger consumers and the industry that serves them develop ever more innovations.

We view this as but one example of an adaptive network of ideas, ratcheting technological developments with opportunities to provide consumers with new values. Demand adapts to possibility, changing the business product environment continuously and radically.

Outline of the Book

Convergence is commonly referred to the synergetic combination of different objects or ideas for other contexts. A revolution of convergence has brought enormous opportunities to develop new products, services, values, business models, and life styles. It has also brought new challenges to knowledge sharing, collaboration, competition, and innovation. Convergenomics refers to the new business environment created by the convergence revolution. This book describes how the megatrends reinforce the whirlwind of convergence which brings both disturbance and new levels of global opportunity. Chapter 2 will present major megatrends in greater detail. In Chapter 3, we will discuss the convergence revolution including technology, digital, industry, knowledge and finally bio-artificial convergence. Chapter 4 will address important developments in e-globalization of business. Chapter 5 introduces the concept of new technological development based on biotechnology, neural science, and nanotechnology. Industry change is considered in Chapter 6, to include digital technology. Chapter 7 describes molecular technology, another area of new development. Chapter 8 considers convergence in the context of supply chains, with focus on lean technology. Chapter 9 will describe the opportunities provided by more interactive use of the internet, to include service oriented architecture, a new technological development that provides great opportunities to work differently through computer software. We will describe how these opportunities provide better access to enterprise systems on the part of small business operations. Chapter 10 focuses on the strategic impact of innovation resulting from the convergence of technology and business needs. The book concludes with Chapter 11, synthesizing the material presented in the book.

References

1 T.L. Friedman. *The World is Flat: A Brief History of the Twenty-First Century*. New York: Farrar, Strauss and Giroux, 2005.

2 M.E. Porter and E.O. Teisberg. *Redefining Health Care: Creating Value-Based Competition on Results*. Cambridge, MA: Harvard Business School Press Books, 2006.

2 Megatrends

The world has always been undergoing change. The average global temperature is cyclical—there are ice ages followed by warming trends. Nature plentifully supplied timber which served as a source of heat for millennia, followed by coal powering the industrial revolution, and by petroleum that serves as the fuel for our primary means of transportation today. While nature appears bountiful, it is fragile and not unlimited. Scarce resources have served both as the motivation for wars in the past and a barrier for population explosion.[1] In the 1960s, world population growth was out of control. Today, many countries face the crisis of population decline. Past excesses in human production have spewed untold tons of pollutants into the air, blackening the skies and purportedly affecting the temperature cycle.[2] Mankind's search for power to drive its commerce has threatened system stability.[3] In the 1950s, nuclear power was thought to provide a safe, inexpensive and non-polluting alternative to fossil fuels. However, nuclear proliferation, especially when the threat is by small and fractionated elements rather than monolithic superpowers, has opened a Pandora's box, creating problems far more dangerous than the serious illness it sought to cure.[4]

The failing energy supply is not the only dilemma we face. Our world provides ample food supplies overall, on land and sea. However, we have a distribution problem. Some areas of the globe generate plentiful food supplies while other areas starve. Moving surpluses from the plentiful places to the starved places is certainly physically possible. However, it is a very complex process to decide upon the appropriate prices, both for energy and food. (And geopolitics plays a major role in allocation decisions.) We have seen economic systems come and go, but the negotiation of prices is a fundamental human dilemma. Another staple of life that is scarce in specific locations is water.[5] In the United States, water shortages are increasing in heavily populated areas (especially growing population centers like Los Angeles, Phoenix, and Las Vegas). China is also facing water shortage over extensive areas. Water is a source of conflict in the Nile, Mesopotamia, and Colorado.[6] Again, we have a distribution problem.

Other problems arise from scientific advances. Complications arise from biotechnology, a field that has seen some of the most impressive advances in the past few decades.[7] Americans have wholeheartedly grasped the increased crop productivity that results from genetic engineering. Europeans and some Asians are far more skeptical. This fundamental issue is still being debated. Biomedicine also promises advances in the control of human maladies, with similar threats from unintended consequences.[8]

In short, the world is complex. There are many issues that have evolved from past solutions. Charles Perrow blamed overly complex human systems for unexpected catastrophes in nuclear energy, petrochemicals, airways, and seaways.[9] In this book we focus on convergence. There has been, and will continue to be, tremendous progress in technology: energy, food, medical, and many other technologies. These are converging, leading us into a world of change. Digital technology also has seen tremendous progress, providing computer technology and electronic tools to make our lives more entertaining and productive. Industry has revitalized itself using these technologies, to provide goods and services more efficiently. Finally, there is a knowledge convergence, with computers generating masses of data from satellites and from chain store outlets into large database storage facilities, available for high-volume data analysis leading to faster, more thorough understanding.

Globalization

Globalization is a term used to describe the development of a connected world, that is much more open and interdependent. This interdependence is made possible by a combination of technological capabilities (it is easy and almost free to communicate globally) and political structure (the end of the cold war politics which has led to a more supportive climate for international cooperation). Globalization has evolved rather than appearing out of nowhere. Everyone knows about the dramatic success of Japanese industry after World War II. A successful manufacturing system was destroyed during the war, but it took very little time to restore itself into the premier quality manufacturing location in the world in the 1970s. Japan has led the way in developing lean manufacturing, the application of quality systems, and other supply-chain management successes.

This tremendous success turned into something of a crisis in the last 15 years.[10] Yet while the cost of manufacturing in Japan suffered relatively, stifling the rate of growth experienced earlier, this seems to represent the maturing of Japanese manufacturing into a niche of high-value products. In turn, there was tremendous growth by Singapore, South Korea, Taiwan, and Hong Kong, the four Asian Tigers. Each of these "tigers" had their own characteristics just as Japan. But all involved systematic coordinated efforts by national firms and multinational corporations in leading Asian progress from 1985 into the 21st century. All four tigers incorporated their own flavors of efficiency in their outwardly oriented economic development.

There are a number of important points we would like to make about globalization:

1. The rate of increase in global trade is about four times the rate of increase in global income. Therefore, we must realize that globalization is driving the economy, not the other way around.

2. Globalization has evolved as many regional economic blocs have developed. These include ASEAN in south-east Asia, the Organization of American States in Latin America, NAFTA in North America, the European Union, and even the Comintern states of the former eastern bloc. Global economic organizations such as the General Agreement on Tariffs and Trade (GATT), the World Trade Organization (WTO), Organization for Economic Cooperation and Development (OECD), Asia-Pacific Economic Cooperation (APEC), G5, G7, G20, and so on, have developed.

3. Globalization became a reality after the fall of the Berlin Wall in 1989, and the end of cold war politics.[11]

4. Globalization is a reality, like the waves of the ocean, and we cannot turn back. However, many developing and underdeveloped countries have opposed globalization, and this has manifested in protests at WTO meetings in Genoa, Italy; Cancun, Mexico; Seattle, Washington; and

> more recently in northern Germany. What is needed is not less globalization, but more equitable globalization.

We now focus on specific contributions in various portions of the world.

Russia

The Cold War provided Russians and Americans with a highly competitive environment, with a generation growing up thinking very little of the other side. In the United States, cartoon characters like Boris, Natasha, and Fearless Leader represented a relatively harmless view of a deep-seated antagonism that expressed itself in its worst form through McCarthyism. Russians were raised to despise the money-grabbing, drug-ridden and corrupt society that espoused rock and roll over humanity. After directing vast chunks of their tax revenues in a never-ending spiral of defense spending, the United States apparently raised the ante too high in the 1980s. Gorbachev decided that the game was not worth the candle, and introduced a period of Glasnost. It made more sense to cooperate than to compete. While neither side has totally overcome four decades of cultural programming, there is less fear of massive nuclear holocaust from the two great powers blasting each other (and the rest of the world) into oblivion.

The Soviet Union is no more having been replaced by the Commonwealth of Independent States (CIS). Russia is the core of the CIS, but the vast resources of the former Soviet Union are shared by a number of new governments seeking to establish themselves. The Iron Curtain also came down, releasing the former Warsaw Pact countries to independence, many of them joined NATO (North Atlantic Treaty Organization) and all of whom are seeking to join the European Union (EU). The turbulent economic change has led many leading scientists and researchers to emigrate. The United States has benefited a great deal from the higher level of mathematical training provided by the former Soviet Union and its former Warsaw Pact allies. The rapid economic growth of Russia under a new economic environment has led to new political confidence in recent years.

European Economic Union

Europe was the focus of the world economy prior to World War I (and to a less degree between world wars). World War II saw massive destruction in Europe, Russia, Japan, and China, while the US manufacturing system exploded. Creative destruction has been cited as potentially positive, and each of these economies has had the opportunity to improve (with obviously different timings). In the 1950s NATO and the Marshall Plan, the European Coal and Steel Community, the Treaties of Rome, Euratom, and the European Economic Community (EEC) arose,[12] leading to overall relative growth controlled at different rates in different countries. In 1987 the Single European Act unified the market more. When the Cold War essentially ended in 1989 with the collapse of the Berlin Wall, the EU was poised for greater regional cooperation, enabling the countries in Europe to participate more effectively in the world market. However, challenges are being faced because of the increasing organizational complexity as more nations join the EU. Anglo-Saxon, Continental, Mediterranean, and Nordic groups have different agendas. There also are different languages, and with each added member of the EU, communication becomes more complex (currently there are 23 official languages that all EU official documents must be translated into for consideration).

China

China has long had the largest population in the world. It was the most advanced country in the world until the 17th century. The Chinese sought isolation for centuries, the desire for which was impinged upon by the British, French, Portuguese, Germans, Americans and others in the name of trade. The vast cultural achievements of the Chinese people remain at least partially a mystery to most outsiders. The Boxer Rebellion and other activities seeking to regain control over their own society were only partially successful. Subsequent Japanese intrusion is still an issue in China. While the United States, rather naïvely, considered all Chinese as allies in fighting the Japanese, they ended up supporting the non-communist factions, which locked in diplomatic positions throughout the Cold War. Most Americans mistakenly think that Chinese nationalism is currently manipulated by Communist Party propaganda.[13] There is strong, and understandable, Chinese desire to be free of domination by external forces.

Chinese society developed on its own after World War II. Some 30 million people had starved due to the misguided Great Leap Forward by 1962.[14] This was followed by the Great Proletarian Cultural Revolution in 1966, lasting almost a decade. This traumatic period saw disruption in educational progress, and relied on a fractionated economic system with little subsequent progress. However, when Deng Xiaoping was in leadership in 1978, he encouraged economic change, using the shocking statement: "It is glorious to be rich." Economic and technological development were encouraged (while retaining control of civil society). Deng had encouraged economic development with "Chinese characteristics." China is still undergoing an economic transition. While we see the most phenomenal economic rates of growth ever experienced, 53 million Chinese were still employed in State Owned Enterprises (SOEs) in the period 1996–2001.[15] There is a complex set of transitional factors going on in China, to include intensifying urbanization (much *ex officio*) to evade relative rural poverty, leading to massive levels of internal immigration. There is also a frenzy of financing, in a relatively uncontrolled banking system, supporting a frenzy of competition as China taps its intellectual capital to produce anything anyone else wants at lower costs enabled by this developing labor system. While politically it remains under the control of the Communist Party, economically China is experiencing fantastic rates of economic growth as a "socialistic market economy."

China is also seeking greater access to the rest of the world (although on its own terms). China sends thousands of students abroad for advanced degrees. Many of those have not returned to China.[16] However, the authors have observed a systematic program on the part of China to attract leading expatriate Chinese researchers back, with considerable success, the birth of the so-called "sea turtles." Chinese economic growth was 400 per cent from 1980 to 2000, the fastest growing economy in the world.[17] This breakneck growth rate of almost 11 per cent per year continued in 2007. This statistical abstraction is clearly understood by looking at the commodities we purchase in the US. It also has had extensive impact on the global economy. The North American Free Trade Agreement was expected by Ross Perot to lead to massive job transfer from the US to Mexico. While Maquiladoras had some growth in the 1990s, the Mexican government has claimed a loss of 218,000 manufacturing jobs from the closure of 500 maquiladoras to Chinese competition between 2001 and 2003.[18]

While the Chinese economy is booming, its consumption is growing at a similar rate. Mao Tse Tung encouraged people to have large families, increasing the 1949 population of 550 million to about 1.3–1.5 billion today.[19] A one-child policy was instituted in 1979, which has stabilized population growth. However, it has also created a new generation of self-centered young people. Brown reviewed the Japanese reliance upon the sea to satisfy their response to growing food needs. Should China do the same, Brown estimated that they would need the fish catch of the entire world! Fortunately for the fish population, China is relying upon increases in consumption of pork and beef. China has chosen to rely upon automobiles for transportation, creating a serious growth in world demand for oil. China now imports more oil than any country other than the US.[20] China had an economic rate of growth of 10 per cent, inducing a demand for an additional 1 million barrels of oil per day.[21] In 2004, Chinese added over 14,000 automobiles per day to their developing road network.

India

The Indian sub-continent is exceeded in population only by China. It has a long history, which in recent centuries has been closely related to Great Britain (for good or for evil). While Indian nationalism is well-developed, the Indian economy has drawn upon English culture for an official language, and for a democratic tradition. Religious differences have played a major role in creating flash points for strife. The conflicts between Islam and other religions have become much more widespread. These conflicts did not begin in the Indian subcontinent, but in the 20th century took their clearest form there.

Despite being vastly overpopulated and culturally fragmented, India has made great strides in contributing to the global economy. In the 1950s, the Indian government took the planned effort to develop a strong computer science educational system. Today, the Indian Institute of Technology system produces some of the finest computer science graduates in the world. In the cultural arts, Bollywood is rapidly generating competition for Hollywood. There was a time when the best and the brightest in India emigrated to the UK, the US, the Soviet Union, and elsewhere. The internet has made it possible to collaborate on software development around the world. There has been a massive development of software engineering firms in India, with many of these firms obtaining level 5 certification from the Carnegie Mellon Institute in order to obtain credentials to verify their software engineering expertise.

The Muslim World

The Muslim world has existed for over 13 centuries. The interactions with the Christian world have involved serious friction throughout history, although for the most part both sides have been able to develop their own cultures, with a great deal of positive cross-influence. While the Crusades led to high degrees of unnecessary violence, for most of these centuries both sides have benefited a great deal from the other.

The West was instrumental in the development of a complex but effective system of transportation based on petroleum. The Muslim world has benefited from control of much of the globe's oil supply. A massive shift in economic power has resulted. Economic cooperation at the top level does not seem to be that big a problem. Middle Eastern states have great petrochemical leverage, motivating their cooperation in the global economy. However, clearly there are fundamentalist forces at play, both in the Muslim world and in the Christian world, that have led to a near *impasse* in cultural understanding.

We see the globe consumed by competitive threats between these not dissimilar societal groups. We only hope that greater communication and contact will bring the world back to its senses. However, recent terrorism has greatly curtailed the closeness of these relationships. Meanwhile, the price of oil is soaring to unimagined heights, although it has become more reasonable since late 2008. We believe this instability will lead to an emphasis on alternative transportation and energy systems.

Thus, we see that major change has happened in many areas that play key roles in our global economy and society. Air travel and the internet are only two factors that draw these diverse developments together. While some players apparently want to be alone, to be isolated from the rest of the globe, that won't happen. Our world is converging.

Digitization

We are in the midst of an information technology revolution. The development of a digital world does not follow a plan, but is the result of a global market, with many participants using internet connections for their own purposes. Ironically, the internet was originally developed as a ploy by the defense research agency DARPA with the intent of providing a communications network immune to blocking by the USSR.[22] The internet has indeed proven to be quite uncontrollable. It has been used to exchange research around the globe, its original purpose. It has been used to support e-mail, allowing people to communicate at a level we could not imagine 30 years ago. It has been used for business, leading to e-business and global supply chains. It is also used as the infrastructure to develop new products, services, and business models.

In the future, there will be even more sharing of software, through service-oriented architecture. It also has been used for nefarious purposes, with internet porn, and gambling, and so on, and as a means for those outside of the law to communicate and do business.

The beauty of the internet is that it is almost beyond control. It is market driven. For some products, this means that greater information sharing leads to more efficient markets. Before the internet, term life insurance was very overpriced. Term life insurance is a very straightforward commodity, offered by many insurance companies, the type of product that is differentiated primarily on price. In the Spring of 1996, Quotesmith.com appeared on the internet, the first of several websites quoting dozens of term life policies (and many other products).[23] The price of term life insurance plummeted. The same thing could be done with many other products, such as automobiles. However, the more other factors are involved in making it attractive to physically touch products, the less impact the internet has. But people are getting more comfortable with accessing products on-line, and that impact will grow.

People use the internet for what they want. This is possible because of its self-healing design, intended to allow communications to find alternative paths when blocked. Some degree of control is possible, as in China. It also is possible to monitor web traffic, to impose after-the-fact control. But a primary characteristic is that it isn't designed, it evolves as the market finds useful.

Another revolutionary digital development is enterprise information systems (EIS, also known as ERP for enterprise resource planning systems). EIS has become a major software product line. An idea started by SAP (originally a German software company) in the early 1970s has evolved into a major information system software product line, which has revolutionized how many large organizations approach business computing. Sales figures are widely available on the internet, but these sources are usually based upon marketing literature which is less than completely accurate. However, the basic picture is clear. Initial arguments were for completely integrated systems, yet vendors usually made sales in the form of modules, covering only limited functions of an organization's computing needs. This was because companies usually were concerned with the high price tags involved, and wanted to minimize their risks by trying out part of the EIS system.

EIS vendor sales peaked in the late 1990s, driven in part by concerns about Y2K problems. This induced many large organizations to adopt EIS as a way to kill two birds with one stone—cleaning up and integrating their organizational computing services at the same time that they assured themselves that they would be Y2K compliant. After that pre-Y2K rush, sales dropped. Vendors then shifted gear, seeking to fill in missing modules in large company systems, and develop products more attractive to small to mid-sized firms. Additionally, vendors have made great strides in reducing some of the trauma of implementing an EIS, making it possible to install systems much faster (a matter of months rather than years), and offering more sophisticated functionality, such as customer relationship management (CRM) and e-business system support. Furthermore, EIS is being marketed heavily in nonprofit sectors, to include both government and educational sectors. Vendor survival seems to depend on the ability to adapt to new market realities, which will continue to evolve. The market is complicated by application service providers. Specialists in software have existed almost as long as software. IBM used to emphasize software support. Firms like Electronic Data Systems provided expertise in particular types of applications. Application service providers (ASPs) now provide expertise on EIS which they rent out to clients, who may either be too small to afford a full functioning EIS or who may not be able to operate them as efficiently.

The web is the backbone of a number of digital advances. Of course, e-commerce itself has been revolutionary. The web has affected supply chains, and those who master it (such as Wal-Mart and Dell) can reach vendors world-wide, creating very profitable supply-chain linkages. The web is key to recent versions of EIS as well. Almost every EIS vendor is now focusing on web portals, which makes obvious sense in the new linked global business environment. An extension of this idea is to use the web to obtain specific digital applications in a services oriented environment, where not only the most appropriate (cost-effective) vendor world-wide can be reached, but within IT operations themselves the best bit of software code for a particular application can be utilized. Advances in digitization have enabled technological convergence. Progress is evident in areas such as wireless mobile devices, broadcasting, biotechnology, the use of radio-frequency identification (RFID) tags, and similar developments.

Changing Demographics

The population of the world is very large, and now stands at around 6 billion people.[24] But the world's demographics are difficult to predict. Drucker reviewed the American experience, which found booms in certain periods, as well as birthrate busts in other periods (1925–1935; 1961–1975).[25] The replacement rate of 2.2 live births per woman would lead to equilibrium. But the actual population is very dynamic. Furthermore, birth rates fluctuate by region of the world. China is very concerned about their birth rate, and has taken action to control it. Drucker reported that most developed and former communist countries have low birth rates. Regions with great problems often have very high birth rates.

The United States has benefited from what has been viewed as the most open, high quality educational system in the world. Our scientific community benefited a great deal from the influx of scientists from Germany in the 1930s. Students from all over the world (especially India, South Korea, Taiwan, and more recently China) have flocked to US graduate schools. This has provided the United States with many of its leading scientists, physicians, and researchers of all types. The positive side of this is the US's gains from having an open society. The negative aspect is that one reason immigrant graduate students have taken so many leadership roles in the US education sector is that those born in the United States seem ever less interested in technical subjects. Top graduates in many US high schools have Indian, Korean, Chinese, or Vietnamese names, exceeding their population proportion. This is great, reflecting strong educationally oriented family systems. But it is disappointing that native Americans don't seem to care about hard academic work.

In fact, we have observed more and more students with Indian, Chinese, and Vietnamese names who speak excellent English, indicating longer times spent in the US, every bit as lackadaisical in their approach to their studies as others with names like Smith or Jones, for example. Friedman reported that many US CEOs feel that outsourcing is attractive, as they get more effort for lower wages.[26] We think this phenomenon reflects an affluent society, where our children thankfully have the opportunity to do many things. We probably need to be patient, and thank the powers that visiting graduate students choose to remain in the US, at least for a while. We also have benefited from a very strong influx of Russian mathematical and computer talent.

One of the reasons so many Chinese graduate students are found in the US is that many of the thousands sent abroad for degrees have stayed. Johnson claimed that this reduced the meritocratic population of administrators in China, retarding their competition with Japan, South Korea, Taiwan, and Singapore.[27] There were 50 million Chinese Communist Party members, but only 2 million of these were college educated. The demographics of China will undergo major changes due to the one-child policy. In the United States, the extended life expectancy of baby boomers will soon tax their descendents in supporting the social security system. It appears that this problem may be magnified in China in forty years or so. In the meantime, there has been a noticeable imbalance in males over females, and some have warned of the pampered generation to come in two decades.

Since Johnson published his work in 2000, the Chinese economy has improved dramatically, and the Chinese government is actively recruiting expatriate Chinese graduate students (now graduates) to return (sea turtles), with some success. We do not know what is the cause and what is the effect between the improved Chinese economy and the increase in repatriation. Regardless, both seem significant developments. These dynamic factors will make human resource management (HRM) very important in China's future, as they need to retain aging knowledge workers, and will have to make decisions about retirement ages. This is because they will face a declining base workforce to support larger retirement pools (much like the United States faces in the near term).

The dramatically falling birth rates and the rapidly aging population in some of the developed and rapidly developing countries present formidable future challenges. Already in Japan, the 60 plus age group represents more than 20 per cent of the population. The shrinking younger generation cannot support the older generation at the current level of quality of life. Japan can hire "invited workers" to fill jobs but the current immigration law will not accept these workers as future Japanese citizens. This double-edged sword phenomenon presents serious social, political, and economic problems to many other countries such as Germany, Spain and South Korea, and so on.

Changing Industry Mix

Change is pervasive in the US economy. Before World War II, the US economic system needed change as most Americans suffered through a traumatic depression. The massive productive capacity created to wage World War II was expected to lead to a post-war slowdown, but saw instead a massive economic consumer explosion. During this period, labor experienced its peak. There has been a steady transition away from human labor to machinery since the 19th century that has eroded labor's influence since the 1960s, when a shift in US industry away from the northeastern rust belt to the south began.

Automation has had a big impact on US jobs. In the 1960s and 1970s, Japanese manufacturing represented the apex of the use of machinery, much of it numerically controlled or computerized which led to greater consistency and quality at less cost. In the 1990s, software systems came into their own, with enterprise resource planning (ERP) replacing many human jobs with efficient integrated systems.

The trends in the 21st century led Forrester Research to estimate the loss of over 800,000 US jobs by the end of 2005 to outsourcing, with another 3.3 million lost by 2015.[28] Other forecasts support the general idea that 10 per cent of the US labor force will be displaced, despite an actual increase in US manufacturing output due to automation. Clearly the US and Western Europe are undergoing transitions in what they do to make a living. In the 19th century the US was predominantly agricultural. This gave way to the manufacturing sector, which dominated the first half of the 20th century, slowly giving way to a more service-oriented economy.

The 21st century promises to emphasize knowledge-intensive industries. We would point out that economic change has been present throughout US history, albeit at a much faster pace in recent times. World-wide, the same transitions have occurred, possibly with different timing. By now, however, the focus on knowledge-intensive industry appears everywhere, especially in developed countries. For example, among the 2007 Fortune Global 100 Companies, 66.7 per cent of US firms are in service industries while 33.3 per cent are in manufacturing. The same trend is seen among EU firms: 61.9 per cent are in services while 39.1 per cent are in manufacturing. In short, values and wealth are being created in such knowledge-intensive industries as biotechnology, pharmaceuticals, ICT, financial services, education, entertainment, and so on. rather than in agriculture and manufacturing.

Deregulation and Privatization

Business firms usually compete best when the market is free of artificial barriers. Thus, many governments have tried to eliminate as many unnecessary regulations and rules that inhibit fair competition as they can. Also, in the same general spirit, many countries began privatizing infrastructure industries to induce innovation. Regulation causes artificial distortion of markets. New technologies work more efficiently in deregulated environments, where competition rewards efficiency. The 1980s saw a wave of deregulation in the United States, radically changing entire industries. Two major industries reflecting this impact were the telephone industry (with the AT&T historical monopoly transforming into many competing telephone companies at the local level, long-distance carriers, and cellular companies) and the airline industry (freed from regulated prices to competitive pricing, creating "yield management").

In the telephone industry, the result of deregulation led to a very high level of competition in the most populated areas of the US. These firms bombarded the public with complex plans, each marketed to emphasize their own relative advantages, yet including enough features to maintain attractive revenue streams. The evolution of technological means of telephone-service delivery includes the internet utilizing free transmission (for the phone companies and users). An early feature of telephone deregulation was "churn," as competing phone plans robbed each other of customers seeking the best deal of the month. This has led to the application of data mining and customer relationship management (CRM) to estimate the value of specific customer profiles to phone service deliverers, and more rational management of the products and incentive programs offered.

American Airlines was an early innovator in the use of software to implement "yield management."[29] Anyone who flies experiences yield management if they compare ticket prices, or have the opportunity to give up their seat because of overbooking. Yield management provides managers with a tool to offer a portfolio of flexible prices for the commodity being sold (such as airline seats on a specific flight), utilizing data about demand, price elasticity, and costs of providing service. Managers in many industries need to establish pricing policies maximizing revenue from a given inventory of items with fixed deadlines. Yield management is applied in many industries, including the hotel industry,[30] railways,[31] car rental,[32] real estate,[33] mortgage lending,[34] and even by government corporations managing natural resources.[35] Specialty software implementing yield management offers a way to differentiate available inventory so that revenue is maintained with less safety stock and thus higher profitability.

Deregulation brings us economic advantage through the lower prices provided by a competitive market. The disadvantages that come with deregulation include much more complex product offerings (as in the telephone market or airline ticket prices), and lower service levels in more remote areas. Furthermore, there always are unscrupulous business people who would like to take advantage of loose rules. The current global financial crisis may be partially attributed to too much deregulation in this industry.

Privatization creates a motivation for profit, introducing a motivation for efficiency. Bureaucracies were formed to provide trained expertise and unbiased, fair judgment in applying regulations though as with many of the best laid plans of humans, it didn't always turn out that way. The communist idea was that altruism would provide best for society. The former socialist or communist countries had only public enterprises—even restaurants, galleries, and supermarkets were State Owned Enterprises (SOEs). Now, developed countries are trying to privatize most functions formerly thought best served by central governments, such as utilities, or infrastructure industries. The current thinking is that the self-interest of the profit motive may yield better results. Privatization provides motivation for people to do their best in return for at least a share of the proceeds of their efficiency.

Commoditization of Processes

The life cycle of innovation makes every new product/service like a toaster, a commodity bought based on price. Advances in technology have made this process even shorter. One good example is software. The internet enables the linking of cooperative software developers throughout the world. That option has made software development projects much more economical. The same idea can be extended to the operation of developed software. The idea of service-oriented architecture (SOA) is to utilize fundamental elements for developing applications.[36] IBM has focused their WebSphere Business Integration scheme on providing a platform where e-business solutions can be integrated.[37] This SOA is a recent extension of the flexibility provided by the internet, which has enabled outsourcing of programming for at least a decade. The result is a much expanded market, where skills can be accessed world-wide.

The internet provides a market in which specific skills can be reached. A service-oriented architecture allows tapping a variety of special software products to perform specific tasks, linked together by grids of computer systems. Outsourcing has become popular as processes are standardized and often certified. Processes such as back-office operations, X-ray interpretation by nighthawks, tax preparation (US income tax code application accomplished in India), and customer service departments are all routinely outsourced.

New Value Networking

Value chains are a relatively recent phenomenon. John D. Rockefeller exemplified the traditional vertically integrated business model, with the intent of retaining every last nickel of profit from drilling petroleum through transporting it to refineries, refining it, packaging it, and selling a variety of products to consumers. Vertical integration sought monopoly. Similar systems were applied to steel and aluminum.

This business model is no longer popular. The current thinking is that organizations should specialize in what they do best, working with others in integrated chains to gather raw materials, transport them, process intermediate forms of materiel, manufacture products, and deliver them to consumers. One exemplar of this approach is Wal-Mart. Wal-Mart gathers many vendors to deliver goods at low cost to consumers. At one time, Sam Walton emphasized buying American. Global realities now find Wal-Mart dealing with a world-wide value chain.[38] Another example of a value chain is Dell, focusing a large set of global suppliers to deliver computers made-to-order.[39] Dell also is undergoing transition in light of economic realities, but Wal-Mart, Dell, and Toyota demonstrate how supply chains of various types can provide greater value to consumers by tapping global partners. The key areas of operations, new customer value, and generating new market base are improved through B-web or E-engineering, drawing upon electronic business.

Emerging New Economies

The world's economic state has constantly been dynamic. Human commerce has transitioned from the Stone Age through other ages such as Bronze, Iron, and so forth. The pace of change has, however, picked up. The Industrial Revolution hit us somewhere in the 19th century, culminating in the English Midlands, the German Ruhr, and the American Rust Belt. Every introductory economics text shows the decline of agricultural employment in the US (less than 2 per cent of jobs), with a corresponding rise in manufacturing. More recently, of course, manufacturing has dropped significantly (less than 14 per cent of jobs). Every change is traumatic, because people have to change what they do for a living. Given the investment we all have made in our professional development, we usually view this negatively. The degree of negative response can vary from violent Luddite-style reaction to joy at the thought of release from drudgery and monotony. It is almost inevitable that new innovations will disrupt the prevailing business system and the global economic scene. But we need to adapt in order to survive.

The global economy is led by four hands: the big hand (the US), a hand (the EU is plugging along without making too many waves), small hands (Japan and the Asian Tigers of Singapore, South Korea, Taiwan, and Hong Kong), and a new hand (China and India). What is especially noteworthy is that newly emerging countries no longer play only supporting roles to firms in the developed countries. Already South Korea and Finland, lead the world in electronics, India is the best source of software engineering talent, Singapore the most progressive economy, and China the fastest growing and now the third largest economy in the world behind the US and Japan. BRIC (Brazil, Russia, India, and China) countries are surging as regional economic powers.

Perhaps one developing economy which deserves our special attention is China. While it has a heavy baggage of its political, social, and economic problems, its influence is enormous in the global economy due to its sheer size and speed. China already has the largest foreign exchange reserve in the world (over US$1.3 trillion) and lent more than US$800 billion to the US government (treasury notes). Many Chinese firms are no longer just outsourcing targets of Western firms. They are becoming major global players and have shown the ambition to absorb some leading US and European firms through hostile take-overs or through mergers and acquisition.

Environmental Concerns, Global Warming

The use of petroleum on a large scale has of course generated enormous pollution problems. The air in Los Angeles is not always safe to breathe. Sometimes air pollution levels in cities like Beijing and Mexico City become dangerously high. Rainwater, once considered pure, now erodes buildings. Many people in Europe and the United States are aware of this, and have tried to educate the world about the need to reduce the use of fossil fuels.[40] The Kyoto Accords are an effort to control the human impact on global warming. However, this becomes a political issue between the "haves", who view the environmental movement as a way to preserve things of value, and the "have-nots". The Third World includes many other nations that look forward to growth that benefited Europe and the US in the past.

There are clear indications that things are warming up.[41] Glaciers in South America are disappearing, as are some islands in the Pacific, and Greenland melts at a much faster rate than normal. Some of this may be due to natural cycles, but the pollution from fossil fuels clearly has some impact as well.

While the physics of the matter are relatively obvious, the political implications are more complicated. There is a growing trend against the Americanization of culture. There is value in retaining unique cultural heritage, and the world does not need to adopt all American ways to progress. Currently severe cultural conflict exists between the West and the East, exemplified by terrorism arising in response to perceived Westernization, Western reaction to terrorism, Muslim reaction to this Western response, and so forth. We can only hope that this seemingly unending spiral can be reconciled and means can be found to accept globalization, and gain its benefits, while retaining cultural purity.

Cultural Conflicts

We live in a world of continuous change with inevitable conflicts. While there are clearly evident struggles over oil, and political stress between Muslim and western cultures, there also are massive changes in pursuit of competitive economic advantage changing the world's entire production system. In this book we describe how information technology, in the form of the web, continues to evolve and impact how business is conducted. These changes are very important, leading to a more open economic world, offering opportunities to many who previously had been isolated and constrained from participation. However, the networked global economy tends to change—some claim even contaminate—unique local cultures. Thus there is a growing tension between those who advocate the global economy and those who strive to preserve their traditional values at all costs. The changing global environment is going to make many who had enjoyed relative prosperity work harder to maintain the style of living to which they have grown accustomed. But these changes are needed to provide greater progress and opportunity to more people in the world.

Convergence Revolution

It is very difficult to maintain high levels of growth. As US growth settled in the 1970s and gave way to Japan as the fastest growing economy, and as Japan gave way in the 1980s to the Four Asian Tigers, there were massive political changes. The Cold War ended, releasing Russia and its satellites to participate in world economic development. Also, China has taken over as the fastest growing economy, while India tries to enter. There have indeed been major changes in many regions of the world in the past decade.

Convergence is a phenomenon that reflects human innovation. The convergence of information technology with computer technology has led to the mobile technology revolution, where executives can Blackberry® through most of their plane travel and teenagers can page and text rather than talk. The digital revolution has led to ubiquitous computing, promising access to enterprise resource planning systems on PDAs (personal digital assistants). Radical changes in demographics will lead to merging of cultures, inevitably leading to sources of conflict, but also promising many gains due to diversity, sharing of cultures, and the generation of new ways of looking at things. Similar changes are occurring in industrial structure. There seems to be a growing trend toward specialization, made practical by the web, which makes it possible to send files containing documents and data around the world in very little time at very little, if any, cost. Human-artificial systems are also making it possible to overcome some of the limitations humans have been confronted with in the past.

Conclusion

We have seen major trends in recent times. Globalization has made the world smaller. We all benefit from our increased access to the rest of the world, broadening our horizons. This aids our cultural development, but also is beneficial economically. Globalization allows us to develop our specific talents to share with the rest of the world. Of course that means that we need to maintain efficiency improvements to remain competitive. Digitalization harnesses computer technology and software applications to our lives. Changing demographics play a major role. The United States has benefited from its traditional role as a melting pot. There are—and have been previously—times when the US has restricted immigration, but we argue that its openness benefits it through the influx of brain- and work-power at all levels, leading to demographic change. Similar change (and resistance) has occurred in Europe.

There also is a need for changes in industry mix, shifting to less polluting means of production, relying in great part on efficiencies from automation. Regulations constrain the economic freedom needed to benefit from economic relative advantage. A climate of deregulation fosters the effective operation of global digitalization. We are experiencing major changes, just as we always have. New economies are emerging, taking advantage of new relative advantages. The world's economy is fragile, depending a great deal on petroleum at the moment, threatening the environment. Different interests also create the potential for conflict, some drawing upon cultural differences. There is a clear change in the economic scenery, with needs for new developments in energy, medical systems, and food. There is a convergence revolution underway, with ramifications that we cannot predict. We need to learn to cope with these changes, and learn to deal with issues, and to take advantage of new opportunities. We now turn to major forces leading to convergence.

References

1 M.T. Klare. *Resource Wars: The New Landscape of Global Conflict*. New York: Henry Holt and Company, 2001.

2 P. Hawken. *The Ecology of Commerce: A Declaration of Sustainability*. New York: HarperBusiness, 1993; Spencer R. Weart, *The Discovery of Global Warming*. Cambridge, MA: Harvard University Press, 2003.

3 L.R. Brown. *Eco-Economy: Building an Economy for the Earth*. New York: W. W. Norton & Co., 2001.

4 G. Allison. *Nuclear Terrorism: The Ultimate Preventable Catastrophe*. New York: Henry Holt and Company, 2004.

5 L.R. Brown. *ibid*.

6 M.T. Klare. *ibid*.

7 P. Pringle. *Food, Inc.: Mendel to Monstanto – The Promises and Perils of the Biotech Harvest*. New York: Simon & Schuster Paperbacks, 2003.

8 F. Fukuyama. *Our Posthuman Future: Consequences of the Biotechnology Revolution*. New York: Picador, 2002.

9 C. Perrow. *Normal Accidents: Living with High-Risk Technologies*. Princeton, NJ: Princeton University Press, 1999.

10 M. Castells. *End of Millenium*, 2nd ed. Malden, MA: Blackwell Publishing, 2000.

11 F. Fukuyama. *The End of History and the Last Man*. New York: Free Press, 1992.

12 *Ibid.*

13 C. Johnson. *Blowback: The Costs and Consequences of American Empire*. New York: Owl Books, 2000. – review of Cultural Revolution, subsequent “It is glorious to be wealthy.”

14 *Ibid.*

15 T.C. Fishman. *China Inc.: How the Rise of the Next Superpower Challenges America and the World*. New York: Scribner, 2005 (2006).

16 C. Johnson. *ibid.*

17 L.R. Brown. *ibid.*

18 T.C. Fishman. *ibid.*

19 *Ibid.*

20 P. Roberts. *The End of Oil: On the Edge of a Perilous New World*. New York: Mariner Books, 2004.

21 T.C. Fishman. *ibid.*

22 M. Castells. *ibid.*

23 S.D. Levitt and S.J. Dubner. *Freakonomics.* New York: William Morrow, 2005, p. 66.

24 D. Meadows, J. Randers and D. Meadows. *Limits to Growth: The 30-Year Update.* White River Junction, VT: Chelsea Green Publishing Company, 2004.

25 P.F. Drucker. *Managing in the Next Society.* New York: St. Martin's Press, 2002.

26 T.L. Friedman. *The World is Flat: A Brief History of the Twenty-First Century.* New York: Farrar, Strauss and Giroux, 2005, p. 260.

27 C. Johnson. *ibid* p. 152.

28 M. Castells. *ibid.*

29 B.C. Smith, D.P. Günther, B.V. Rao and R.M. Ratliff. "E-commerce and operations research in airline planning, marketing, and distribution," *Interfaces* 31, No. 2, March–April 2001, 37–55.

30 C. Dosh, Corps. "Face scarce space," *Business Travel News* 23, No. 4, 3/6/2006, 23–31.

31 Anon. "A route to higher profit," *Railway Gazette International* 162, No. 3. March 2006, 130.

32 J. Zhu. "Using turndowns to estimate the latent demand in a car rental unconstrained demand forecast," *Journal of Revenue & Pricing Management* 4, No. 4, January 2006, 344–53.

33 E. Johnson. "Yielding results," *Journal of Property Management* 71, No. 1, January-February 2006, 43–7.

34 C. Brinn. "Improving yield management," *Broker Magazine* 7, No. 5, August 2005, 68–9.

35 R.M. Nazzaro and T.S. Hephner. "Valles Caldera: Trust has made some progress, but needs to do more to meet statutory goals: GAO-06-98," *GAO Reports* 11/16/2005, 1–47.

36 M.P. Papazoglou and D. Georgakopoulos. "Service-oriented computing," *Communications of the ACM* 46, No. 10, October 2003, 24–8.

37 K. Bhaskaran and M.-T. Schmidt. "WebSphere Business Integration: An architectural overview," *IBM Systems Journal* 43, No. 2, 2004, 238–54.

38 C. Fishman. *The Wal-Mart Effect*. New York: Penguin Books, 2006

39 S. Holzner. *How Dell Does It: Using Speed and Innovation to Achieve Extraordinary Results*. New York: McGraw-Hill, 2006.

40 L.R. Brown, *ibid*; Hermann Scheer, *The Solar Economy: Renewable Energy for a Sustainable Global Future*. London: Earthscan, 2002, orig. in German 1999.

41 P. Hawken, *ibid*; S.R. Weart, *ibid*.

3 Convergence Revolution

There has been a fundamental change in how business is done in the 21st century. The driving force for the change is convergence in a number of areas. Science continues its progress, although each advance seems to bring with it concerns about control. Information systems have converged allowing global exchange of data, information, and knowledge. Technology has advanced allowing greater compatibility between components produced across the globe. These convergences allow cooperative efforts for business and industry systems throughout the world. The data generated by compatible systems and organizations support a convergence in knowledge development and exchange. Finally, opportunities are being explored in the biological and computer science areas, leading to many complex developments that hold promise for a better quality of life.

Evolution of Convergence

Convergence is a result of human creativity, usually obtained by taking ideas from one area of experience and applying these ideas to other areas. Thus, convergence has been part of human history. In business innovations, convergence has also evolved through several major stages as shown in Figure 3.1.

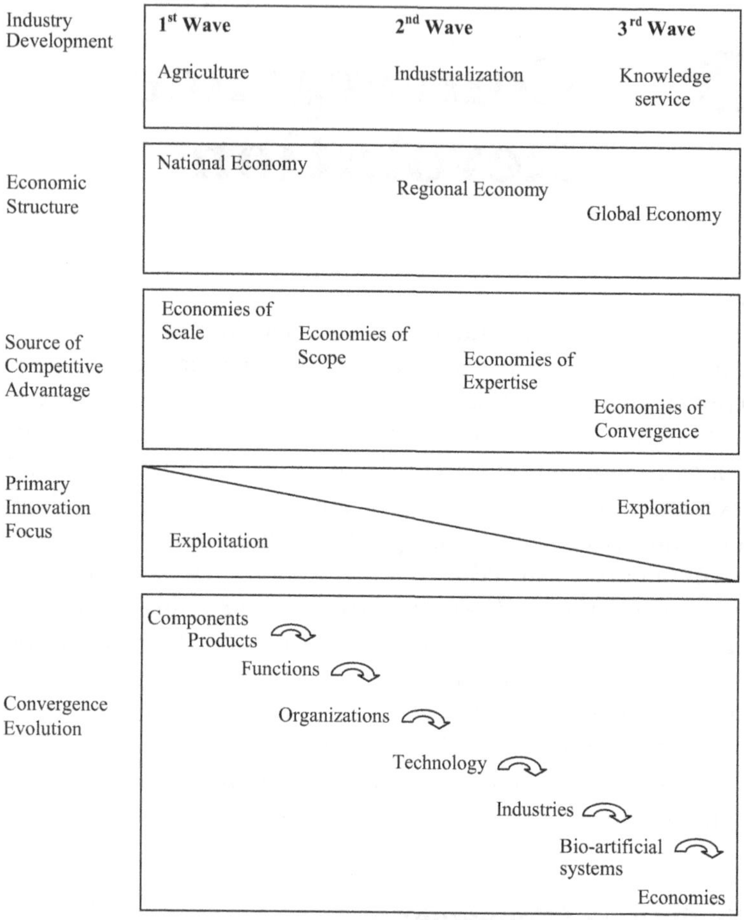

Figure 3.1 Evolution of business environment, strategies and convergence

Organizations' efforts for innovation through convergence were influenced by several layers of major factors:

1. megatrends;

2. change in industry mix;
3. change in economic structure;
4. change in the source of corporate competitive advantage; and
5. the primary innovation strategy.

Humans are creative when they make connections, absorbing diverse perspectives and ideas.[1] The ability to connect ideas across areas has grown over cultures (globalization), knowledge (digitization, technology integration), industries, economic boundaries, strategic competitive advantage, and primary innovation strategy. The driving force of the economy in advanced countries has transitioned from agriculture to manufacturing and to knowledge-intensive service industries. The economic emphasis has also changed from national to regional and now to global economy. The source of competitive advantage has progressed from economies of scale, to economies of scale and scope, to economies of scale–scope–expertise to new economies of scale–scope–expertise–convergence. The primary innovative strategy has gradually gravitated from exploitation of current competencies to exploration of new competencies. The success of any organization is largely based on its ability to predict the future direction of convergence that can enrich the entire value chain—operational innovation, new products/services, new customer values, and new customer base. Evolution of convergence can be examined in six broad levels, from components/products, functions, organizations, technology, industry, and biological/artificial systems, that are relevant to organizational innovation. Figure 3.1 presents evolution of the business environment, corporate strategies, and convergence. Table 3.1 summarizes the convergence levels and their basic purposes to create organizational innovation.

Table 3.1 Evolution of convergence and organizational innovation

	Convergence Level	Purpose
Level 1	Component/product convergence	Product innovation
Level 2	Functional convergence	Process innovation
Level 3	Organizational convergence	Value chain innovation
Level 4	Technology convergence	Technology, product innovation
Level 5	Industry convergence	New industries, customer value innovation
Level 6	Bio-artificial systems	Ubiquitous innovation

Component/Product Convergence

Convergence reflects human intelligence in putting things together in new ways. It has always been with us, and can be as simple as combining an eraser with a pencil, placing the eraser on one end of a lead pencil for obvious convenience. Things like adjustable wrenches represent a form of convergence, giving mechanics the ability to deal with any size of nut rather than having to have maybe 20 different sized wrenches. In more recent times, convergence appears in the form of an electronic toothbrush, converging the time-honored need to clean teeth, a brush with electronics to charge a battery that assures we brush our teeth the full two minutes our dentist requires, and a more thorough and consistent motion to the brush to assure it can get into those hard-to-reach places. For children, the electric toothbrush is combined with Disney characters to make fun to brush teeth more regularly. Component/ product convergence can provide an organization with a new competitive advantage through product innovation.

Functional Convergence

A process is a logical set of related activities taking inputs, adding value through doing things, and creating an output.[2] The traditional organizational structure is based on vertical semi-independent functional areas such as manufacturing, sales, finance, human resources, and so on. These silos had vertical walls and thus cross-functional processes were not very efficient. In business, there are many different ways to get work done. Information systems play a key role. They provide a means to collect data, store it efficiently, generate reports to let management know what the organization is doing, and archive data for future reference as needed. Operational processes have to do with accomplishing cross-functional value creation, including product development, order management, and customer support. Infrastructure processes are more administrative, such as establishing and implementing strategy, and managing many aspects of the organization to include human resources, physical assets, and information systems. Each of these generic processes, whether operational or relating to infrastructure, involves sets of tasks needed to accomplish work. For example, in the operational process of order management, it is necessary to forecast the volume of demand expected for the products produced by an organization. The function of forecasting can be accomplished in many ways:

- using last month's demand as a prediction for this month;
- using the monthly demand from a year ago as the prediction for this month;
- applying a spreadsheet algorithm such as exponential smoothing over available monthly data;
- incorporating seasonality indices into such a spreadsheet algorithm;
- taking known orders and adjusting forecasts based on past demand records;
- relying upon managerial judgment;
- using Ouija boards (throwing darts; rolling dice; guessing).

Business process re-engineering (BPR) is an activity in which an organization accomplishes process innovation to make the value chain more efficient by destroying functional walls. BPR is closely tied to enterprise information systems (ERP, for enterprise resource planning), because in order for ERP to provide benefit to the organization, at least some of the ways in which that organization does business must change.

Hammer and Stanton noted that re-engineering has sometimes been a euphemism for mindless downsizing.[3] They gave an example of how re-engineering should work. In the early 1990s, Texas Instruments faced long cycle times and declining sales. BPR was applied, with calculator development accomplished by cross-disciplinary teams from engineering, marketing, and other departments. These teams were to be in control of every aspect of product development, from design through marketing. The first pilot teams failed, sabotaged by the existing functional departments, which felt threatened. Functional departments were unwilling to give up good people, space, or responsibility. Power continued to lie in the old functional departments. Texas Instruments responded by changing the way it was organized. Development teams became the primary organizational units. Functional departments focused on redefined missions supporting the product teams. Budget was accomplished by process instead of by department. Office space was re-allocated. After the new system became established, new product launch-time was cut by as much as one-half, and profitability was enhanced, with return on investment multiplied fourfold.

Organizational Convergence

Economist Ronald Coase theorized that firms will expand until the costs of organizing a marginal transaction internally equal the costs of conducting the same transaction in the open market.[4] In terms of value chains, each organization will bring in its core competence to maximize value creation, shifting from vertical integration (internal operations) to horizontal integration (value networking).

New technological developments have changed the way organizations are structured. Communications technology has made it possible to combine the skills of people in different locations. When a firm had to physically send specialists to problem areas, this was a constraint on the physical structure of organizations. Either an organization had to be large enough to hire specialists in each location where they were needed, or they had to rely upon transportation systems to send them to problem areas. Electronic communications release organizations from such constraints. Telephones helped, but were limited to voice mode. The internet enables richer communication in both reach and richness of information, with e-mail to provide written text, and file transfer to allow more complete document sharing. This break of the trade-off between reach and richness of information enables global networking. It is no longer necessary to move an Indian computer engineer from Hydrabad to New York City to gain their services. The engineer can stay in Hydrabad, remaining in his or her own traditional culture, paid much less for his or her time, than they would be in the US, but able to live as well or better. The same idea extended to subcontracting leads to outsourcing, and at an even higher level, the ability to create global supply chains. Service oriented architecture (SOA) and web services liberate businesses from their rigid legacy systems, enabling an on-demand business model.

Additional convergence is made possible across countries. The internet has been key in allowing the exchange of digital information globally. This has enabled real-time exchange of computer files, which quickly expanded into exchange of work. The latter half of the 20th century saw the development of more effective means of production, moving away from the old view of efficiency that stressed making more things cheaper to an approach that sought to minimize waste. Waste was minimized by only producing what was necessary, in addition to producing it well.

Not only can organizations spread out throughout the globe, technology enables linkage of cooperative organizations into supply chains, which have enabled organizations such as Wal-Mart in retailing and Dell in personal computer manufacturing to improve their performance. Value chains reflect a focus on contributions of each supply chain member toward delivering better products to consumers, thus accomplishing value chain innovation.

Technology Convergence

The 1990s saw a great deal of convergence in the technology area. Information technology boomed, with the power of computers growing by leaps and bounds (Moore's Law). Communications technology also saw rapid development, and this led to a convergence of information and communications technology enabling devices such as personal digital assistants (PDAs), evolving into the ubiquitous Blackberry®, MP3 music technology, and a continuing growth of new and better ways to communicate and transfer information. The 1990s also saw the development of bio- and nano-technology, leading to biopills. Other developments in medicine saw the application of laser technology in medical treatments. Technology convergence creates new technologies and product innovation.

These advances are expected to bring better things, in food production, medical technology, and energy. However, the ramifications of these advances are expected to continue to be perplexing, bringing with them new dilemmas, just as coal power, nuclear energy, and petroleum have in the past. Convergences of technology lead to human creation of better ways of doing things, in the form of nanotechnology, better products and services, neuroscience, biotechnology, and other fields advancing progress.

Another example of technology convergence is digital convergence. Digital convergence is a consequence of ubiquitous computing, also referred to as invisible or ambient computing, coined by Mark Weiser of Xerox in 1991. Digital convergence involves analog–digital integration, wired–wireless integration, voice–data integration, and service–device integration, which result in convergence of networks and telecom-broadcasting convergence.

We've already talked about the convergence of people linked by the web. There are two other major factors in digital convergence. Enterprise information systems (EIS, and ERP) have become very important in modern business operations.

Even as early as 1998, over 70 per cent of Fortune 1000 companies had either begun implementation of EIS, or were planning them.[5] These systems have been credited with reducing inventories, shortening cycle times, lowering costs, and improving supply-chain management practices. Enterprise information systems are designed to integrate all of an organization's information system computing. They have been credited with increasing the speed with which information flows through a company.[6] Enterprise information systems have also been credited with:

- creating value through integrating networks across a firm or firms (Metcalfe's Law);
- the implementation of best practices for each business process;
- the standardization of processes within organizations;
- one-source data resulting in less confusion and error; and
- on-line access to information.[7]

All of these features facilitate better organizational planning, communication, and collaboration. Applied Robotics increased on-time deliveries 40 per cent after implementing their ERP, and Delta Electronics reduced production control labor requirements by 65 per cent.[8]

Industry Convergence

Convergence has occurred in many business and technological sectors, including the telecom and financial industries. The telecommunications and computer industries have seen powerful digital components brought to market at ever lower costs (Moore's Law). The business applications of developing technology led to the replacement of analog telephone systems with fully digitized networks. This has evolved into a continuum of products that could be classified as either telecommunications or computer products, with the subsequent merging of industries.[9] Duysters and Hagedoorn cite the convergence of the telecommunications and computer industries into a single information and entertainment industry. We can think of the *Star Wars* movies as a manifestation of this convergence. Chon *et al.* cited a similar convergence in the news media-telecommunications industry, which they attributed to the digitization cited earlier along with deregulation.[10] There is also convergence within the financial services industry, with banks and insurance companies cross-selling products.[11] Industry convergence can create new industries and customer value innovation.

Convergence can lead to gains in other areas as well. The convergence of technology with opportunities in the communications and music industries have led to the iPod®. Convergence of technological capability, educational opportunities, and the dispersed location of people has led to the development of distance learning, utilizing media in the form of recording technology, communication technology for two-way (and multi-way) communication, file sharing technology for learning materials, and web-based education. Gains in healthcare were obtained by treating all processes as a series of ongoing experiments, which is accomplished by measuring and comparing results with standards. This step also requires that all processes be spelled out in detail to avoid ambiguity. Problems identified are immediately addressed. Rather than working around problems, their causes are identified, and countermeasures established. Successful methods are disseminated after assuring that it is appropriate for each process. Experimentation is encouraged at all levels of the organization.

In addition to cutting waste and improving quality, the new industrial paradigm includes linkage in the form of supply chains, often global in scope. A variety of systems have evolved to coordinate the diverse players in global supply chains. Supply chains are networks of suppliers/vendors, manufacturers, distributors, and retailers that are connected by transportation, information, and financial infrastructure. The idea of a coordinated supply chain manifested itself in the form of vertical integration in the 19th century, leading to consolidation of ownership and interlocking centrally coordinated organizations. These mammoth business operations proved unwieldy, however, in the faster-paced business operations of the latter 20th century, where the ability to specialize and coordinate through e-business has led to complex supply chains that seem to develop from markets rather than the plans of specific people. Business brought many advances in the management of supply chains. Current supply chain technology includes potentially many suppliers/vendors linked to one or more manufacturers and/or assemblers, who often distribute through multiple distributors to usually many retailers. Industry supply chain leaders such as i2 and Manugistics have developed tools to support sharing of sales and forecast data, and enterprise resource planning software often feature such support.[12]

Biological and Artificial System Convergence

The march of technological progress continues, as it has for centuries. Norman Borlaug was recognized for developing wheat that was resistant to stem rust. This advance promised better yield of crops in areas of scarcity, which was labeled the "green revolution." The expectation was that overpopulated and under-producing areas would be able to sustain themselves. In the 1990s, one of the most popular areas of technological growth was in genetic engineering of food. However, by 1998 the Japanese and Europeans had rejected what they referred to as "Frankenfood."[13] In the United States, genetic engineering has enabled much more cost-efficient food productivity. But the US also has a minority of consumers convinced of the value of natural, organic food. In Europe and Japan, this is a controlling majority. While further scientific progress is expected to continue in many fields, it is also expected that debates will follow each of them.

There is similar technological advance in the medical field. Fukuyama[14] proposed three possible futuristic scenarios in this crucial branch of science:

1. Drugs such as Ritalin and Prozac aid self-esteem and concentration, but have side-effects for specific types of individuals. Functional foods combining focus on nutrition and health could change the entire life science sector, converging the food and pharmaceutical industry.[15]

Genomics in the future may enable drugs to be tailored to genetic profiles (medical anthropology), minimizing side effects. This might provide a means for individuals (who can afford the drugs) to specifically attain psychological moods on demand without risk of addiction, hangover, or brain damage.

2. Stem-cell research can regenerate almost any body tissue, providing physicians with tools capable of extending life well beyond 100 years for those who can afford it. However, humans are complex systems, and other aspects of aging may make such gains moot. Further, extended life may clog social systems (minor ones being Social Security and Medicare in the US).

3. Human reproduction may be supported by technological advances allowing prospective parents to design their offspring in terms of intelligence and/or appearance, as some Chinese parents practice for their only child. This may sound attractive, but deeper reflection scares many people. Germans in the 1930s were pioneers in this area of biological study and the Chinese are making advances in this area today.

The idea that computers can evolve into life-forms is interesting to some. Ray Kurzweil discussed the potential of computers to gain intelligence in multiple books.[16] Others have discussed the anticipated convergence of information technology, biology, and business.[17] The former focus was on building organizations to predict and control, to sense change and to respond. Kurzweil has even predicted that by about 2040 we will reach a singularity, where computers attain the power to function as well and even better than humans. He finds that attractive. We aren't so sure. Molecular technologies enable bio- and nano-technology, with the potential for greater connectivity, utilizing autonomous software to converge information, biology, and business to achieve ubiquitous innovation.

The evolution proposed by Kurzweil, leading to ever-smarter computers capable of functioning at or beyond the human level, can lead to a greater understanding of molecular technology environments. This will lead to many new opportunities to develop radically new and better products as well as processes to get things done. Ubiquitous computer environments are emerging in our own time. This can, however, get out of hand. The danger is that we may reach the state of singularity where humans lose control to artificial systems. Kurzweil sees no danger in this, allowing computers to take care of mundane activities, freeing humans to presumably enjoy life in whatever manner they wish. We are not so sure that this would be a good thing. However, with the possibility of a singularity and the associated advance in computer potential, computers will be ever more ubiquitous, part of practically every aspect of human life. Some of this is already possible in monitoring the state of machinery and appliances, providing the potential for automatic fault detection and self-correction. The application of computers to aid in health care of individual humans offers another field of great potential benefit.

Convergence of Economies

The highest structural convergence would be convergence of economies, such as the EU. There are also other types of cooperative or collaborative convergence, such as OPEC, ASEAN, NAFTA, WTO, the G5, the G7, OECD, Free Trade Agreements, and so on. We have seen an evolution in political structure. Fukuyama traced progress from tribal-based organizations depending upon agricultural subsistence and slavery, replaced by better systems with theocracies, monarchies, and feudal aristocracies.[18] The United States is based upon a reaction to these centralized forms of political structure. Communism was another reaction. We now see modern liberal democracy dominating in the political domain, and technologically driven capitalism in the economic domain.

Many economic agreements have formed over the past four decades. Practically every area of the world has a trade association such as ASEAN in south-east Asia. Some have disappeared, such as the Comintern (replaced by similar organizations crossing former ideological lines). More recent organizations of this type include NAFTA and the EU. The overall trend is to a more open, free trade environment world-wide, although protectionism still appears on occasion.

Conclusion

We have seen developments that led to a revolution in how the world of business is organized and conducted. Technologically, great challenges exist in food, medicine and energy. We have developed the ability to produce ample food, but have two problems. First is the problem of distribution. Some areas of the world have surplus, others starve. Getting food from one to the other is an age-old problem. The physical problems are easily solved. The economic and political transfer problems are not. Great strides are being made in the medical field, but the system for payment is a complex problem. Energy is an even greater challenge. Fossil fuel energy has served the world quite well for almost two centuries. However, it has created many difficulties in the form of pollution, and there are real concerns about the sustainability of petroleum. A great deal of energy is being devoted to developing alternative forms of energy, but the problem is not at all easy, and the impact on the world is very threatening.

A second area of convergence offers mitigation of problems if not solution. We have made tremendous gains in digital technology, leading to extensive global communication on an individual basis. People can talk, which should help solve easy problems. They also can reach markets, which gives them greater reasons for hope. Organizations have more powerful, integrated tools to aid them in coping with the masses of data they now face. Organizations can also reach broader markets to obtain the digital services they need. They also have more powerful tools to manipulate and interpret the data generated. However, there still remain problems with controlling the internet, dealing with the complexity of massive organizational systems, assuring quality in an open web services environment, and ethical issues in the use of data.

Another area of convergence is the organization of production. Digital technology provides the opportunity to link businesses. Supply chains have emerged linking raw materials to customers through ever-more efficient groupings of organizations. By accessing broader markets of suppliers and vendors world-wide, production organizations have shifted manufacturing work around the globe, impacting every social system. China faces problems in dealing with exploding opportunities in the manufacturing sector versus declining interest in agricultural production. Mexico faces transitionary problems in trying to respond to a once-growing maquilladora system which is being outsourced to China. The US faces complex problems in dealing with a shifting work-force structure with rust belt employment disappearing, heavy influx of Mexican workers causing concerns about immigration, and the need to develop new skills of value to the world from a growing workforce. Similar problems occur in Europe.

Some of the most exciting opportunities involve convergence in biotechnology. This is a sensitive area, as consumers are easily scared away from products if there is the slightest hint of possible problems. However, any technology involves complex relationships that could be potential problems. We have evolved over millennia, with natural selection discarding characteristics with the worst problems, and will continue to do so. New ways of doing things are needed to make life better. Convergence of scientific thinking and needs for improvement will lead to further progress in biotechnology, as it has in other areas.

Overall, we appear to be making quite a bit of progress. We have made impressive gains in food, medical, and energy technology, in digital technology, in supply chain organization, and in knowledge management. However, as has been true throughout history, every gain in human understanding is like opening a door to many more mysteries. The need for greater understanding grows exponentially as we learn more.

References

1 J. Burke. *Connections*. New York: Simon and Schuster, 2007.

2 H.J. Harrington, E.K.C. Essing, and H. van Nimwegen. *Business Process Improvement Workbook: Documentation, Analysis, Design, and Management of Business Process Improvement*. New York: McGraw-Hill, 1997.

3 M. Hammer and S. Stanton. "How process enterprises really work," *Harvard Business Review*, November-December 1999, 108–18.

4 R.H. Coase. "The nature of the firm," *Econometrica* 4, 1937, 386–405.

5 T. Minahan. "Enterprise resource planning: Strategies not included," *Purchasing* 125, No. 1, July 16, 1998, 112–27.

6 T.H. Davenport. "Putting the enterprise into the enterprise system," *Harvard Business Review*, July-August 1998, 121–31.

7 D.E. O'Leary. *Enterprise Resource Planning Systems: Systems, Life Cycle, Electronic Commerce, and Risk.* Cambridge: Cambridge University Press, 2000. For business process reengineering, see M. Hammer and S. Stanton, *ibid.*

8 M. Weil. "Managing to win," *Manufacturing Systems* 17, No. 11, Nov. 1999, 14.

9 G. Duysters and J. Hagedoorn. "Technological convergence in the IT industry: The role of strategic technology alliances and technological competencies," *International Journal of the Economics of Business* 5, No. 3, 1998, 355–68.

10 B.S. Chon, J.H. Choi, G.A. Barnett, J.A. Danowski, S.H. Joo. "A structural analysis of media convergence: Cross-industry mergers and acquisitions in the information industries," *Journal of Media Economics* 16, No. 3, 2003 141–57.

11 L. Van den Berghe and K. Verweire. "Convergence in the financial services industry," *The Geneva Papers on Risk and Insurance* 25, No. 2, 2000, 262–72.

12 K. Xu and Y. Dong. "Information gaming in demand collaboration and supply chain performance," *Journal of Business Logistics* 25, No. 1, 2004, 121–44.

13 P. Pringle. *Food, Inc.: Mendel to Monstanto – The Promises and Perils of the Biotech Harvest*. New York: Simon & Schuster Paperbacks, 2003.

14 F. Fukuyama. *Our Posthuman Future: Consequences of the Biotechnology Revolution*. New York: Picador, 2002.

15 M. Brännback, A.J. de Heer and P. Wicklund. "The convergence of the pharmaceutical and the food industry through functional food: Strategic change and business opportunity or an illusion?" *Pharmaceuticals Policy and Law* 5, 2002, 63–78.

16 R. Kurzweil. *The Age of Spiritual Machines: When Computers Exceed Human Intelligence*. New York: Penguin Books, 2000; *The Singularity is Near: When Humans Transcend Biology*. New York: Penguin Books, 2006.

17 C. Meyer and S. Davis. *It's Alive: The Coming Convergence of Information, Biology, and Business*. New York: Crown Business, 2003.

18 F. Fukuyama. *The End of History and the Last Man*. New York: The Free Press, 1992.

4 E-Globalization Strategy

New Rules for the E-Global Age

As we live in the new e-global age, corporations must learn to take advantage of the many opportunities that new developments in electronic technology provide. Their future success will be determined by how well they are able to harness new technologies in innovative ways, furthering the future corporation's strategies. Leadership of these future corporations will need to be able to sort out threats and opportunities in the new megatrends they face.

Future corporations will need to understand and operate within new rules as they enter the knowledge/intelligence age. Those that are able to innovate, and generate new value from the tools that technology provides, will be able to capture flows of wealth. The e-global age is highly dynamic, and firms are not able to seek optimal solutions as Henry Ford was able to do. Ford was able to optimize his assembly lines, in a stable engineering environment. In the future, we expect that change will be so rapid that firms will have to obtain advantage through adaptive innovation. They will have to make hay while the sun shines, as competition will be able to rapidly enter successful markets.

This will make it necessary for innovative firms to find new products of value for their customers, exploit them while the opportunity presents itself, and move on when competition (and new innovation) will drive the profitability of innovative products down. For every new development, there will be a new set of rules for firms to learn and exploit.

In part, value will be generated by the explosiveness of network growth. Robert Metcalfe argued that the value of a telecommunications network is proportional to the square of the number of system users. In a telephone system, if there is only one phone, the system has no value. It doesn't have too much more if there are only two phones on the network. When millions of telephones are connected, the system is much more valuable. It has been argued that the impact is even greater in virtual communities and group communications created by Web 2.0, as there are exponentially more ways in which network connections can be created. The ability to access millions or potentially billions over the internet makes it far more valuable than a local area network.

The value of international business is explained in part by this phenomenon. It is valuable to e-businesses within the closed United States to have network connections with their customers through the web. We have seen that this can include the ability to not only talk to anyone in the world, to sell to anyone in the world, and to work together on projects, together through service oriented architecture.

A recent IBM study[1] echoes many of the points we have discussed above. This study, based on interviews with more than 1,000 CEOs from around the world, concluded that the enterprise of the future will need the following traits for success.

1. Hungry for change: visions for change, strategies for change management, and processes for new business models.
2. Innovative beyond customer imagination: delight customers with innovative new values, global presence while remaining locally relevant, and maintain deep collaborative relationships with customers.
3. Globally integrated: networked with other global partners for world-class capabilities, developing leaders with global mindset, and seek global innovation.
4. Disruptive by nature: challenge the current business model and rules of competition, create an environment for entrepreneurs and innovative business models, and lead the industry in this effort.

The Network Economy

In 1997, Kevin Kelly published 12 principles of the network economy.[2] These principles were based on four axes.

1. Wealth will flow directly from innovation in the dynamic network economy, not from optimization. Thus participants in this economy should focus on seizing the unknown with imperfections rather than seeking the perfection of optimization.
2. The ideal environment to cultivate the unknown is agility and nimbleness provided by networks.

3. Domestication of the unknown inevitably requires abandoning the perfected known. Participants will have to learn to give up what were good things in the past.

4. As the network economic web thickens, the cycle of find, mature, and destroy will happen faster and with greater intensity.

We have already discussed Metcalfe's Law. Kelly used this concept for multiple principles. First, the law of plentitude—more gives more—networks are more valuable and powerful the more participants it has. And success is nonlinear. An example is Microsoft or SAP, currently atop the information technology profit heap. Both organizations went through long gestation periods with little profitability. Amazon.com was another famous example of a firm that suffered negative profits year after year, but finally reached some tipping point where profits rolled in at phenomenal rates.

In network economies, significance precedes momentum. The epidemiological analogy of a tipping point is when a disease has infected sufficient hosts to allow infection moving from local illness to raging epidemic. Technological tipping points seem to be much lower than they are in biology. This is a major factor in the dynamism of the network economy.

As demonstrated by Metcalfe's Law, value explodes with membership. This leads to the law of increasing returns. The focus should be on looking for increasing returns for the network, as opposed to the old thinking of seeking economies of scale. Those firms that take advantage of these new rules of economies of network will be in a position to prosper, and "those that have shall get more."

Economic pricing is being stood on its head. Transaction costs dive toward the free, as in the telecom business. This leads to the law of inverse pricing. In the past, consumers sought slight improvements in quality for slight increases in price. The microprocessor industry flipped this price equation.[3] Firms must now anticipate that drastically superior quality will be available at lower cost. In the enterprise system market, firms like Compiere offer their software for free. The innovative strategy in this environment is tied to the law of generosity. If the free price is successful and attracts sufficient market, the network of users will explode exponentially. As everyone adopts this technology, those in a position to service it can build a business model based on service and assistance contracting (as is done by Compiere). Microsoft applied the same strategy with their web browser, and Sun did the same with their Java language. (And of course Linux operating systems are the most recognizable form of open software, available for free).

The network economy will not provide only change, but rapid churn. The law of churn topples incumbent firms and creates a platform ideal for greater innovation and birth of new products through sustainable disequilibrium. This is akin to the Hindu concept of Shiva, a creative force of destruction and genesis. The concept is also reminiscent of Schumpeter's views of economic progress through creative destruction.[4]

In the new network economy, our ability to solve social and economic problems will be limited primarily by our lack of imagination in seeking opportunities. Drucker's idea of not solving problems, but rather seeking opportunities, reflects the law of inefficiencies. It implies a refocus away from optimal solutions to passé systems to grabbing on to the wave of the future.

Distruptive and Sustaining Technology

Business strategy will need to consider a variety of important facets of management in new light. By considering disruptive and sustainable features, business managers faced with declining performance may be able to redirect firm strategy to take advantage of new opportunities. For instance, firms losing out to competitors in terms of product cost and quality may be able to survive and prosper by redirecting strategy to speed and customization. Zara was able to utilize information technology in the highly competitive fashion clothing market in Europe. They adopted a revolutionary strategy of producing minimal quantities of models, sufficient to show customers what they had available. Rather than suffer high inventory losses from unfashionable lines, they were able to respond to market demands by ramping up production very rapidly on those lines the market liked.

There are interesting dichotomies in the way in which organizations are structured. John D. Rockefeller, Andrew Carnegie, and Henry Ford believed in vertical integration. If there was any aspect of their industry in which a penny was made, they wanted to control that aspect. US Steel owned coal mines and iron ore mines, as well as mines for the many alloys used in steel. They owned shipping lines and railroads to deliver ores to their mills. These mills included coke plants, blast furnaces, smelters, and rolling mills. Ford Motors used to own steel plants and rubber plantations in Brazil for tire production. Firms such as IBM and General Motors continue to operate in integrated forms.

But vertical integration is inherently inefficient. Organizations have a difficult time being efficient at everything they do, especially when they generate large bureaucracies to administer ever increasing management functions. The e-global age emphasizes shared work and specialization, relying upon value chains of many participants in more open markets. Firms like Microsoft, Intel, and Dell prosper through their value chains. Nike utilizes its network of participants to coordinate all of its production activities, allowing Nike itself to actually produce nothing (except brand image). Information technology enables firms to coordinate, as we will discuss in Chapter 8. Sustaining technical cooperation is clearly winning out over disruptive independent entities.

The e-global age includes ideas such as lean manufacturing and adopting just-in-time principles to focus on inventory reduction and quality. This enables evolutionary innovation of Kaizen, as production is carefully monitored without the clutter of high work-in-process levels. There also have been advances in enterprise information systems, such as provided by SAP and Oracle, among the world's largest software vendors. Business process re-engineering focuses on analyzing the way in which work is done, and usually includes improvements in efficiency and cost through the use of information technology. This is a form of revolutionary innovation that is widely available to firms operating in the e-global age.

Globalization Strategy of the Firm

Successful firms in the global age must be equipped to compete and thrive in the global market. Global strategy is much more than a larger scale of domestic business strategy. There are many challenges and uncertainties in the global market such as cultural differences, language barriers, government laws and regulations, importance of different markets, unique needs of local customers, and the like. Nevertheless, global strategy is imperative for several reasons. First, sustainable growth of the organization requires global funding and expansion as we have seen in GE, Microsoft, IBM, Google, Apple, Infosys and the like. Second, efficiency of the value chain can be dramatically improved through global operations such as global sourcing, strategic alliances, and joint ventures as shown by Nike, Mattel, and so on. Third, today knowledge creation, sharing and application often demand global partners. Fourth, corporate customers are now global, often e-customers who purchase goods or services on-line. Fifth, most formidable competitors are now global firms in other countries, thus requiring global strategy.

Developing a Corporate Global Mindset

Developing and sustaining a global mindset is the fundamental requirement for global strategy. A global mindset implies thinking globally and locally simultaneously. If firms are not open to diversity across cultures and markets, they are not capable of attaining this global mindset. This implies the need to be knowledgeable about diversity across cultures and markets, and the ability to integrate such knowledge into corporate strategy.

A global mindset can only be attained if a core group provides leadership. This needs to be accomplished at the managerial level, and adoption of an organizational culture at lower levels for successful implementation. Top management needs to provide the required structure, process, and empowerment for this culture to prosper.

By developing a corporate global mindset, organizations will be able to operate over the web throughout the globe. The corporate global mindset will provide such organizations with the understanding of local cultural differences, social differences, and understanding of political environments that will give them the ability to generate more appropriate responses to challenges of conducting international business.

The corporate global mindset will be represented by a focus on effectiveness in providing value to customers rather than the 20th century norm of increasing productivity. This will require consideration of efficiency in terms of delivery of value rather than utilization of resources at lowest cost. There also will be a need to emphasize fairness over organizational emphasis on control and local autonomy.

Building Global Competitive Advantage

While the world is getting flatter because of globalization, ICT advances, the end of cold war politics, and so on, it is by no means an open unified market. As a matter of fact, the greater discrepancies in economic and technological advances have also deepened regional differences around the globe. Ghemawat presented a new approach to develop a global competitive advantage, AAA Triangle Analysis.[5] The Triple A Triangle Analysis involves examination of an organization's three global strategy dimensions *vis-à-vis* major competitors:

1. Aggregation: the organization's strategy to seek economies of scale and scope with global consistency (for example, McDonalds, IBM, Microsoft).

2. Adaptation: the organization's effort to maximize local relevance of its products and services (for example, mini-IBMs in each country, Google China).

3. Arbitrage: the organization's strategy to exploit the differences between national and regional markets (for example, headquarters in the US, factory in China, call center in India, retail shops in the US and Western Europe).

The Triple A Triangle Analysis suggests a simplified illustration as shown in Figure 4.1. The main idea is how the organization's global strategy is positioned as compared to its major competitors. The organization may want to focus on one or two of the Triple A dimensions to develop its competitive advantage. For example, it may focus on arbitrage strategy while keeping aggregation and adaptation at about the median level among its competitors. Or, it may focus on adaptation to maximize its local relevance while keeping the other two strategic dimensions at about their average levels. Since there

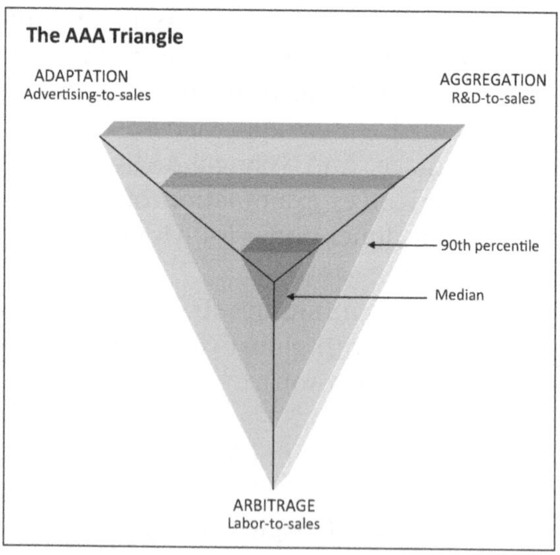

Figure 4.1 The AAA triangle analysis

is no single variable which can represent each of the three dimensions, Ghemawat suggested proxy measures as follows:

Aggregation = Total R&D expenditure/total sales;

Adaptation = Total advertising expenditures/total sales; and

Arbitrage = Total labor cost/total sales.

Global Outsourcing

Operating in the e-global era provides opportunities to work together with organizations throughout the world. Value chains can be created that enable specialization by comparative advantage released from the constraints of location. Or, production can be shifted depending on the changing global environment. Larry Summers, the former President of Harvard University, has described the emergence of developing market countries in playing greater roles in the global supply chain. Summers argues that this participation will expand, as individuals in developing countries will play greater roles as knowledge assets and as employees, and expansion of production organizations in developing countries will generate expanded opportunities through outsourcing or even offshoring. This will lead to the global spread of business, as opposed to the old concentration of production in developed countries. This offers many benefits:

- Transaction costs will be lowered, as value chains can tap the most efficient producers for components of their products. This will require coordinated specifications, but the internet's ability to speedily exchange information will provide vast opportunities to take advantage of operating on a vastly larger working environment (and market).

- This implies vertical disintegration. High levels of vertical integration will gradually disappear, as value chains will emerge providing more efficient and effective coordinated efforts on a temporary basis. This implies a highly dynamic business environment.

- Information and communication technology (ICT) is a key enabler in this new global economy. This will enable development of temporary core competences in a highly competitive environment.

There also will be some features of the e-global age that will be troublesome to many. First, it implies a considerable degree of job transfer. The old vertically integrated economies led to antiquated structures involving a triad of negotiation, often resulting in agreements between labor and capital at the expense of consumers. This resulted in higher product costs, often protected by tariff structures. The ability to escape the physical bounds of such systems through e-business has resulted in the tumbling down of highly inefficient production systems. When this hollowing out of production capacity leads to the reduction of production capacity in basic industries to near zero levels, many governments get concerned about their ability to produce what had been strategically important commodities, such as oil, steel, and aluminum. It might be the case that strategic production capacity may be jeopardized, but we see a much greater overall benefit from a market where shared production based upon competitive advantage is present, with less emphasis on political competitiveness.

The Future Corporation

The future of business environments has been the topic of consideration of a number of leading scholars and corporations. We already have discussed the IBM study earlier in this chapter. Michael Porter of Harvard views the future as one in which needed skills will develop so rapidly that there will be a surplus of individuals with outdated skills, and shortages of specialists with needed knowledge and technological skills.[6] There will be much greater need for individuals with the ability to manage knowledge, requiring both computer literacy and in-depth understanding of e-business requirements, global business, and the technical specialties required. Those having scarce and valuable skills will become free agents, able to select their organizations. Corporations will thus need to devise incentives to attract key knowledge workers. Individuals would be well advised to keep current with developing computer and production technologies.

Nobuyagi Idei, a former CEO of Sony Corporation, viewed convergence beyond that of technologies to convergence of entire industries. For instance, electronic firms such as Sony and Samsung could collaborate with computer firms such as Dell and Apple and production firms (or at least value chain firms assembling products) such as Nike for the development of innovative new products. This implies the need for temporary groupings of people to modularly fit in evolving value chains.

Peter Drucker was probably the most prolific business author in the past 30 years. He expressed his views of the prospects offered to business in the 21st century.[7] He viewed our times as one of profound transition, with more radical change than was experienced in the industrial revolution of the 19th century or those triggered by the traumatic events of the mid-20th century. He outlined five social and political certainties that would shape business strategy:

1. the collapsing birthrate in the developed world;

2. shifts in distribution of disposable income;

3. redefinition of corporate performance;

4. the need to be competitive on a global basis;

5. growing incongruence between economic and political realities.

Drucker saw the new information revolution calling for emphasis on the meaning of information rather than on the technology needed to gather it, and on knowledge-worker productivity with those workers being viewed as capital assets rather than production costs. Knowledge workers were expected to call for them to be responsible for managing themselves as free agents in careers in a highly dynamic business environment. The future value chains will be developed and managed by a federation of partner organizations who bring their own world-class core competencies.

Convergence Examples

We have witnessed dramatic developments in e-business. Three pioneers utilizing supply chain technology enabled by e-business are Nike, Wal-Mart, and Dell. Nike doesn't actually make shoes—they create demand through massive advertising campaigns through star athletes, and fulfill that demand through a network of global supplier links interconnected electronically. The internet provides a means to generate work at efficient locations throughout the world. Dell assembles computers, but like Nike, doesn't make much of their product. They also rely upon electronic connections to link numerous global vendors to provide the components they need. Dell has benefited a great deal in the past from their ability to minimize inventory through their make-to-order system, enabled by internet linkage. Wal-Mart has revolutionized the retail market in the United States, generating complementary competitors around the world such as Carrefours, each adopting a different business model, but each utilizing efficient supply connections to tap efficient vendors globally. Wal-Mart is probably the world's largest private user of data, with its managerial personnel and vendors linked to a comprehensive data mining and decision support system credited with saving them operating expenses in the billions of dollars annually.

Each of these global supply chain businesses are enabled through the convergence of electronic technology with business opportunity. The shoes provided by Nike are not necessarily dramatically different from those of their competitors. Their innovation comes from operating their business more efficiently through global business. Dell also doesn't provide a product different than its competitors, but like Nike utilizes electronic technology to do business much more efficiently. Wal-Mart provides basic products for its customers, but at significantly lower cost than was available prior to their innovation.

The Human Equation

A new approach to working with knowledge workers in the 21st century is needed. Technological equality has truly flattened the world. The old 20th century views implied by TQM, JIT, BPR, and ERP were basically negative, trying to fix what was wrong and viewing labor as a cost to be reduced. The emphasis was on replacing erratic and costly human labor with computer technology.

Humans will be the key in the 21st century e-global age. But a new positive approach will be needed. Management will be responsible to identify employee strengths, interests and knowledge. Individual employees will have a greater onus to keep current, constantly renewing their knowledge, thus enabling them to create more value through greater engagement. Instead of the old adversarial labor/management dichotomy, knowledge workers and management need to consider themselves partners in cooperative efforts that can be linked through knowledge networks, with the shared mission of production to fulfillment of consumer needs.

References

1 IBM. The Enterprise of the Future, Global CEO Study, 2008.

2 K. Kelly. "Twelve principles of the network economy," *Wired* 5 September 1997.

3 C.M. Christensen. *The Innovator's Dilemma: When New Technologies Cause Great Firms to Fail*. Cambridge, MA: Harvard Business School Press, 1997.

4 J.A. Schumpeter. *Capitalism, Socialism and Democracy*. New York: Harper and Row, 1942.

5 P. Ghemawat. "Managing Differences: The Central Challenge of Global Strategy, *Harvard Business Review* 85, No. 3, 2007, 59–68.

6 M. Porter. *Competitive Advantage: Creating and Sustaining Superior Performance*. New York: Free Press, 1998.

7 P. Drucker. *Management Challenges for the 21st Century*. New York: Collins, 2001.

5 Technology Convergence

There have been dramatic developments in science, especially in the areas of information and communication technology (ICT), bio- and nano-technology. The past few decades saw liquid-crystal displays (LCDs) replace clock dials in our offices and bedrooms, and all kinds of meters in our automobiles. LCD technology is advancing to provide us better products because science has given even greater miniaturization of LCDs, brilliantly displaying texts and graphics. This makes smaller computer devices possible, so that we can watch better quality movies on airplane flights or cross-country drives. They make possible more realistic video games, which are entertaining, but also enhance the ability to teach or learn.

While LCDs have been very useful, they could be more powerful than we might like, the images they produce have difficulty fighting sunlight, and devices may be thicker and less flexible than we might want. New developments in this field have led to crisper images, brighter colors, thinner devices, and lower power consumption. E-book devices may replace books and newspapers through use of ambient light from the surrounding area, yielding sharp images readable even in bright sunlight. Once a page is displayed, no further power is needed to keep it visible, as power is only consumed during updating. Newer devices have also been found to be easier on human eyes. This technology has been brought to market by Sony (marketing the Reader) and Motorola (Motofone). In Italy, Telecom Italia is planning to market the Librofonino, an e-book reader with a cellular connection for receiving information.[1]

Biotechnology has received a great deal of interest in the past 20 years. The convergence between the fields of computer technology and biology has yielded many useful innovations. The electroluminescence of organic compounds can be utilized to build organic light-emitting diodes (OLEDs), thinner and brighter than LCDs, with wider viewing angles. These devices have thus far appeared in small devices such as secondary displays on the exterior of mobile phones, or in music players. Lifetime is about 20,000 hours, which typically translates to two years, making them unattractive in laptop computers or televisions. But technology is improving. With additional gains, OLEDs could provide superior brightness, contrast, and response to LCD televisions.[2]

An example of convergence of technology with life needs is provided by Silicon Valley, where the firm Overtime Fitness offers "exergaming" equipment based on video games focused on exercise. They utilize Xbox technology to exercise, aiming to reduce exercise's monotony. This application was probably initiated in 1989 when Nintendo released the Power Pad accessory for its Entertainment System, which gamers could use to translate step on or jump on to compete in their "World Class Track Meet." They also sold a Power Glove controller. Neither was a commercial success at that time. In 1998 Konami released an arcade game, "Dance Dance Revolution," where contenders danced on an electronic platform, competing to correctly follow patterns. This was more successful commercially. It involves so much exercise that the idea was natural to extend the market to include the fitness industry.[3]

The Influence of Technological Convergence on Innovation

Innovation has a life cycle, shaped like an S curve, with long periods of initial effort yielding an invention that, if successful, can rapidly lead to an innovative product with a strong growth rate, followed by a period of product maturity and ultimately demand decline (often due to new innovations). Economist Joseph Schumpeter hypothesized creative destruction, describing a discontinuous rate of economic development as new innovations do not evolve from old developments, but rather eliminate them.[4] More recent theories emphasize a chain of incremental innovations that can be viewed as evolutionary.[5] Wonglimpiyaret demonstrated how developments in Windows operating systems, and the technological evolution from video tape recorders through VCRs and DVDs to new devices follow a sequence of small jumps (breakthroughs) that over time seem like a wave of S-curves in a near-continuous stream of innovation.[6] Innovation in the bank card industry has led to widespread use of debit cards (electronic fund transfer at point of sale). Magnetic stripe technology developments enabled innovations from credit cards (circa 1966) to ATM cash cards to debit cards (circa 1974) to Smart cards, enabling e-cash, mobile phone banking, telephone cards, and pay TV cards.

Biotechnology

In the drug industry, biotechnology draws upon convergence between pharmaceuticals, agriculture, and the chemical industry to apply technology in the form of organic chemistry, pharmacology, biochemistry, and molecular biology to provide new and better medical products. New drugs based on genomics can be viewed as radical breakthroughs.[7] There is the potential to treat patients at the genetic level, a convergence offering designer drugs that can treat individual genetic structures.[8] Wonglimpiyarat described the wave progression from physiological chemistry (circa 1800) to organic chemistry and pharmacology in the 1900s to biochemistry breakthroughs starting around 1985 to molecular biology and molecular diagnostics based on genomic technologies. There is a convergence between biotechnology (bio-electronics, drug delivery) and nanotechnology (electrical, biomedical, energy, and the environment) leading to better tools in genomics, bioinformatics, nano-devices, and information technology.

An example of the convergence of biology with computer technology involves neural technology. Neuroscientists are very interested in the neocortex, as it promises better understanding of how to treat psychiatric disorders and brain diseases, and also could explain how learning and memory work. IBM has cooperated with Swiss research partners to create a functioning three dimensional model of a two millimeter bit of neocortex. The project was labeled "Blue Brain." Raw data was collected from rat neurons and fed to a supercomputer, which provided a graphical representation. This rat neurocortical column contained 10,000 neurons, but took near two terabytes of storage (the same amount of data would need to be held on 400 standard recordable DVDs).[9]

Academic institutions are beginning to see the benefits of convergence between engineering and biology. Such convergences have been widespread since the 1980s, with efforts intended to improve the understanding of cancer and other diseases at both the micro- (sub-cellular) and macro- (population) levels. These efforts can lead to biological computers. Already there have been benefits from such convergence in the form of Magnetic Resonance Imaging (MRI) and Positron Emission Tomography (PET) scanning. Further development may lead to atomic force microscopy, enabling scanning at the cellular and sub-cellular level. Convergence includes application of modeling approaches taken from engineering and applied to biology. Professor Richard Kitney of Imperial College in the UK explained that modeling of cellular networks for mobile telephones has implications for better understanding of biological cells and how they communicate. Multiple feedback loops used in aircraft system design have aided in analysis of blood pressure control systems.[10]

Other gains from the convergence of engineering and biological studies are found at the University of California at Santa Barbara, which initiated a laboratory in 2000 with the aid of Microsoft Corporation. The lab focuses on developing new technologies to enable people to interact with computers in more powerful ways. One output of the effort is HandVu, a vision-based system for hand-gesture interface. Users carry a computer and wear goggles with an attached camera. The computer is activated by sticking a hand into the camera's field of view. The camera locks onto the hand and tracks its movements, displaying the changing scene in the goggles. Various gestures are used as computer commands. For instance, a fist followed by a scissors motion with thumb and forefinger grabs a virtual object in the computer display. The virtual object is released by opening the fist and pointing thumb and fingers forward. Potential opportunities arise if the computer can be linked to global positioning systems (GPS), with applications such as displaying an on-demand map flashing restaurant menus. Hand-tracking computers are a step in the progress of virtual reality.[11]

Neural Science

Neural systems enable us to physically survive, by controlling our sleep patterns, our breathing, and bodily function. Every day, millions are treated through the use of anesthesia. Yet how that puts us to sleep is not well understood. Researchers at the University of Copenhagen have found evidence involving sound waves, an alternative to the traditional wisdom that nerves are activated by electrical currents. Better understanding of the process can improve the results obtained from anesthesia. Other research, at SUNY Downstate Medical Center, involves a hormone which normally has a tranquilizing effect, but in adolescents has been observed to do the opposite, and trigger anxiety.[12]

The addition of practical application to research understanding of human processes is exemplified by subliminal advertising. This concept was popularized in the 1970s, using advertising in forms where the observer didn't realize what they were experiencing. Subliminal ads supposedly worked below the threshold of human recognition. No one knew if subliminal advertising really worked, but the Federal Trade Commission specifically forbade it over public airways. Current brain-science research is finding that some of the principles of subliminal advertising seem valid, and further work is needed to more fully understand how our minds work.[13]

Nanotechnology

Nanotechnology involves the study of entities at very small scales. A nanometer is one billionth of a meter. The average human hair is 100,000 nanometers. Human blood cells are 2000 to 5000 nanometers long. A hydrogen atom is 0.1

nanometers wide. The ability to understand and manipulate matter at the nanometer level would provide much greater ability to understand matter and biological function.[14]

Nanotechnology is active in almost every area of science, including computer science, electrical engineering, and mechanical engineering, promising to develop new knowledge that can be applied to many fields. Biotechnology can benefit through increased understanding of the basic building blocks of living matter. Understanding how information is encoded in DNA can lead to new products through the process of convergence. The genetic basis for specific diseases may allow us to correct harmful mutations. This could lead to the ability to improve existing humans, or design better future humans.[15]

In the field of information technology, Moore's (or Hwang's) Law addresses the phenomena of price reduction, speed, and capacity of computer systems increasing exponentially, accompanied by steady miniaturization. Continued decreases in transistor size are constrained by heat dissipation and electron tunneling. Nanotechnology offers better understanding which could lead to further breakthroughs, which then could lead to better ways to utilize computers to do new things through convergence with application areas.

Computer science is also interested in how the human brain makes decisions. Robotics software has proven beneficial in many arenas, including manufacturing, replacing humans with robots in dangerous environments, or in applications prone to errors. Nanotechnology in the mechanical engineering level can lead to better production machines with longer wear. Nanotechnology offers gains in electronics, life sciences (biomedicine and biopharmacy), materials science (fuel cells), machines, and the environment.

Evidence of Convergence

Product convergence has existed throughout time. We obtained clock radios through the convergence of two mature products. Cell phones represent convergence of functions such as internet access, music access, photography, and even telecommunications. But there are opportunities for many new instances of convergence through technological developments in ICT, biotechnology, and nanotechnology. Greater understanding of how the world works is a mechanism necessary to the development of new and better methods. This convergence is driving many universities to exciting research fields. It is also driving electronics giants such as IBM, Philips, and ST-Micro to join Japanese powerhouses in the fields of biotech, nanotechnology, and microelectromechanical systems (MEMS).[16]

Biotechnology and neural science have been very evident in the development of new knowledge, reflecting society's values with respect to rights to health care. This of course has a major impact on life styles. The creation of new knowledge arising from the convergence revolution is evident in the development of better life style choices as the consuming population becomes more aware of the impact on health of many products and behaviors. A simple example of this is the massive growth in gymnasiums and health centers, a detectable switch away from highly sugared sodas to water (at least on the part of older people).

Efforts in nanotechnology research have yielded benefits. Toshiba recently announced that it could inject nanoparticles into living cells to induce specific reactions. These MEMS devices generate vibrations bonding nanoparticles into a fluid to attach to cell surfaces, and thermal energy to physically alter cellular surfaces to inject nanoparticles into the cells. The MEMS chip is a nanoparticle manipulator with a water-repellent diaphragm.[17]

We have seen astounding progress in scientific research in recent years. Here we have reviewed but the tip of the iceberg in understanding how biological systems function, how our nervous systems function, and the physics of the small can be applied to better tools, processes, and methods. We will next look at how industry has benefited from these advances

References

1 *The Economist.* "Displays to keep an eye on," 10 March 2007, pp. 4, 6.

2 *Ibid.*

3 *The Economist.* "Let's get physical," 10 March 2007, p. 6.

4 J.A. Schumpeter. *The Theory of Economic Development, 5th ed.* New York: Oxford University Press, 1967.

5 G. Dosi, C. Freeman, R. Nelson, L. Soete and G. Silverberg. *Technical Change and Economic Theory.* London: Pinter, 1988.

6 Wonglimpiyarat. "Convergence for innovation," *Asia-Pacific Tech Monitor* September-October, 2006, 16–24.

7 *Ibid.*

8 R. Kurzweil. *The Singularity is Near: When Humans Transcend Biology*, New York: Penguin Books, 2005.

9 D. Talbot. "IBM: The computer brain," *Technology Review*, September, 2005, 52.

10 *The Engineer*. "Life less ordinary," 18 June–1 July, 2007, 30–31.

11 *Convergence: The Magazine of Engineering and the Sciences at UC Santa Barbara*, "Is that tea real?" 7: Spring, 2007, 3–5.

12 K. Capell. "Of neural sound waves and raging hormones," *Business Week* 4028, 4/2/2007, 69.

13 D.E. Schultz. "Subliminal ad notions still resonate today," *Marketing News* 41:5, 3/15/2007, 5, 9.

14 J.V. Kennedy. "Nanotechnology: Breaking through the next big frontier of knowledge," *Indiana Business Review* Summer, 2007, 7–9.

15 D.E. Duncan. *Masterminds: Genius, DNA, and the Quest to Rewrite Life*. New York: Harper Perennial, 2005.

16 K. Gilleo. "The new convergence: Bio, nano and MEMS appear a 'magic mix' of science and technology," *Circuits Assembly* 17:2, Feb, 2006, 32.

17 *Ibid*.

6 Industry Convergence

Convergence involves developments in different fields that can be utilized to do something in a manner better than was possible before the convergence. When Fulton developed a functioning steam engine, ocean travel across the Atlantic was ultimately transformed, cutting travel time from an uncertain number of weeks to a predictable number of days. When Edison came up with the light bulb, city lighting was transformed from smelly, unreliable, and risky nights to the incandescent city glow that makes urban living so attractive. When Edison, who specialized in such things, came up with a moving picture camera, entertainment was transformed by the convergence of technology with the need to reach more people than could fit in a theater. In fact, Hollywood was created by the convergence of Edison's sharp business practices with creative competition, which found the climate much healthier in southern California than it was in New Jersey.

Industry convergence occurs when two or more previously distinct industries become direct competitors or cooperators. For example, when Apple competes with music distributors that is industry convergence; when Walt Disney went into the travel convention business that was a convergence of entertainment, hotel, and golf industries. Malhotra and Gupta give two types of industry convergence.[1] Functional convergence is when customers behave as though the products of two industries are linked. Personal computers and television industries both provide some of the same function, and become substitutes. Complementary convergence occurs when multiple objects of value are obtained in one location, such as getting travel and lodging services in the same location (as when travel agents sell both airline tickets and hotel rooms, maybe even throwing in ground transportation). Institutional convergence is when firms behave as though products of two industries are linked. The banking industry has converged, so that they can provide both checking deposit services as well as mortgage loans. A railroad executive was once credited with the concept that it was a mistake to think of their business as the railroad business—it should be thought of as the transportation business. Some oil industry firms claim that they are in the energy business, allowing them the latitude to develop petroleum as well as hybrid fuels (even though sometimes the motivation is more to control competition).

Convergence in the News Industry

The mobile phone has evolved from the convergence of electronics with telecommunications. Internet technology has transformed many industries including telecommunications. The news industry was transformed when cable networks came along, and CNN and ESPN took over industry dominance from CBS, NBC, and ABC. The internet offers opportunities for a new way for us to access the news. We have had the opportunity to obtain polyglot opinions via blogs. But the web is now viewed as an alternative news medium. For instance, the *Wall Street Journal* is available in both traditional print form as well as on-line.

Associated Press Broadcast Technology is reported to be developing a new medium.[2] This medium has the ability for immediate dissemination of news globally. The news can be accessed anywhere and at anytime, to include use of mobile phones. This opportunity motivates greater integration between broadcast news and print systems. It is driven by the technology on the customer side, due to sharp increases in broadband connectivity and the development of acceptable levels of speed and capability in mobile phones.

Digital convergence is also for IT and broadcasting. Bill Burke, product manager for ENPS of Associated Press Broadcast Technology, sees a shift in focus from output at given intervals (such as the evening news) to pools of content distributable in multiple ways. Editorial work will be unified to serve all media, providing output for television, radio, print, and the web. This will change reporting assignments. There will not be a regular work day—news will occur in real time. The focus will shift to create packages of raw information and media links that can be subsequently shaped into content that is structured for each media type.

This will require a common database repository of news that is accessed by a common communications protocol. This protocol will allow production elements to exchange information and pull material from the database. Work in progress can remain in the database and be updated as events evolve. A distribution layer will allow each media to draw content. A television producer might create finished content for a mobile phone newscast, but that might be only rough-cut content for web delivery. One of the benefits of such a system is consistency gained from use of a common database.

Telecom Industry Convergence

The 1996 *Telecommunications Act* split the American Telephone and Telegraph (AT&T) monopoly into a variety of firms, to include seven regional companies. One type of industry convergence is merger and acquisition (M&A), and this has been found in the telephony industry, where only one of those seven operating companies (Bell South) continues in its original form. The other six have evolved into three companies (SBC, Qwest, Verizon). One path was the purchase of NINEX by Bell Atlantic, and the subsequent merger of Bell Atlantic with GTE to become Verizon. US West merged with long-distance and fiber-optic company Qwest. BellSouth has also been active, forming Cingular with SBC.[3] Mergers continue, and Cingular has become the new AT&T. While we do not expect a complete return to a monolithic AT&T of the 1980s, convergence is clearly occurring at the corporate level.

While one cannot hope to trace all of the ownership arrangements involved in mergers and acquisitions, probably more interesting is the evolution of industries in telephony. There are at least four types of company organizations in this field:

1. network providers (both landline and wireless);
2. tool providers (hardware and software);
3. transaction and service providers (buying and selling over the internet; other internet services);
4. internet/content providers (copyright material distributed via the internet, or through other channels).

This industrial reorganization is a convergence in our terms because it combines new technology with business opportunities. Satellite technology has obvious impact through making wireless networks possible. Software and hardware tools directly impact tool providers. The internet and the developments that make it easier to use drive transaction and service providers, as well as content providers.

Industry Use of Digital Convergence

Digital convergence is the unification of functions, combining technological development with business opportunity. Digital convergence provided great strides in delivering better consumer goods, to include on-demand television, epitomized by TIVO and competing products, musical downloads (for good and for bad), movies over the web, and many other services related to business which appear to be accessible through service-oriented architecture. We are constrained by business models, as in the music business, where many details continue to be worked out to fairly share the proceeds of the industry. Regulations also are present, although issues related to what can and shouldn't be available over the internet are under debate in the United States, China, and the rest of the world.

Digital convergence has been present for decades. Semiconductors, computing, communications, and other aspects of electronics provide tools at a very rapid rate.[4] Over that period, we have seen tremendous changes in industry. IT has been relied upon to do more and more. One indication of this growing role is the percentage of IT investment of total investment by US business, which was 2.6 per cent in 1970, 3.5 per cent in 1980, 9 per cent in 1990, and 22 per cent in 1999. This proportion of total investment dropped slightly after Y2K, but is still obviously of growing importance in doing business.[5] A more recent study of US productivity based on data through 2004 found evidence that IT investment intensity has some positive impact on productivity acceleration.[6]

A branch of computer technology development offering a great deal of promise is mobile technology. Wireless Application Technology (WAP), Bluetooth (wireless communication standard), and third or fourth generation (3G, 4G) mobile phone systems allowing broadband access vastly improve the potential for users to access more of their computer tools and data wherever they want (or nearly—wherever they have wireless access). There has been a rapid growth in nomadic computing, using computing in much broader contexts through mobility, mass scale, and digital convergence.[7] South Korea has embarked on WiBro service and 4G systems.

Yet another area of digital convergence is found in the use of the internet. Internet service providers (ISPs) provide content. ISPs need to invest in equipment and telecommunications access. Many large application service providers (ASPs) exist (AT&T WorldNet, IBM Global Network, AOL, MCI, T-Online) that own their own telecommunications lines. This involves a type of convergence itself, as telecommunication and media companies are converging with content providers, such as the merger of AOL and TimeWarner.[8]

Another aspect of digital convergence is the development of standards through the market.[9] Microsoft markets Smartphone/Pocket PC, which competes with products such as Symbian OS, Palm OS, J2ME by Sun, and Linux OS. Standards are not enforced by one party on the rest of the world. Instead, standards emerge via compromise of mutually competing networks. A sort of digital convergence emerges meaning that consumers have the capacity to access any kind of content using any type of terminal (PC, mobile telephone, television) at any time and at any place.

Evidence of Convergence

The rate of growth in information and communication technology has been phenomenal. We have seen how this has led to much better market offerings for consumers in both the IT and news industries. There are similar gains in many other industries. A precondition to this growth is technological development, which we saw in the section on digitization. The future of business belongs to those who can take advantage of the opportunities they find in converging these technologies to do more things in better, more efficient ways.

Convergence is a term used in the telephony industry to denote the provision of multiple functions within one service. An example is a system linking traditional telephone connections (wire) with cellular connection over IP connections, and wireless fidelity (WiFi) connection over fiber and/or coaxial cable lines.[10] This enables use of wire and wireless networks with a single user identify, and provides a mechanism to obtain many personalized services, such as voice activated programming of digital video recorders, video conferencing, and other telephone capabilities.

The auto industry also has converged with digital technologies.[11] Digital devices are available to not only give turn-by-turn directions, but that can alert us to traffic jams and suggest alternative routes. Jaguar was planning an add-on entertainment system with music, photos, or movies stored on flash cards, as well as the widely available satellite anti-theft service. While these services have been available on luxury models, they are dropping in price, enabling their use in more affordable vehicles. Microsoft is developing navigation software for a wide variety of models with many add-on products. Fiat was working on wireless connectivity for the road.

These are only two areas in which new products and services derive from industry convergence. Greater collaboration across disciplines has often led to greater innovation, driven by competition, such as edutainment (education and entertainment) or infotainment (information and entertainment).

References

1 A. Malhotra and A.K. Gupta. "An investigation of firms' strategic response to industry convergence," *Academy of Management Proceedings*, 2001, G1–G6.

2 "Convergence makes a comeback," *IBE*, January-February 2006, 36–37.

3 V. Grover and K. Saeed. "The telecommunication industry revisited: The changing pattern of partnerships," *Communications of the ACM* 46, No. 7, 2003, 119–25.

4 D.B. Yoffie. "Competing in the age of digital convergence," *California Management Review* 38, No. 4, 2006, 31–53.

5 C. Rettig. "The trouble with enterprise software," *MIT Sloan Management Review* Fall, 2007, 21–27.

6 C. Corrado, P. Lengermann, E.J. Bartelsman and J.J. Beaulieu. "Sectoral productivity in the United States: Recent developments and the role of IT," *German Economic Review* 8, No. 2, 2007, 188–210.

7 K. Lyytinen and Y. Yoo. "Research commentary: The next wave of nomadic computing," *Information Systems Research* 13, No. 4, 2002, 377–88.

8 S. Magin, R. Algesheimer, F. Huber and A. Herrmann. "The impact of brand personality and customer satisfaction on customer's loyalty: Theoretical approach and findings of a causal analytical study in the sector of Internet Service Providers," *Electronic Markets* 13, No. 4, 2003, 294–308.

9 I. Oshri and C. Weeber. "Cooperation and competition standards-setting activities in the digitization era: The case of wireless information devices," *Technology Analysis & Strategic Management* 18, No. 2, 2006, 265–83.

10 D. Schwartz. "Mobilizing for mobility: Delivering the 'quadruple play' with fixed mobile convergence," *www.cedmagazine.com*, September 2005, 30–1.

11 D. Tynan. "Cars that are smarter than you are," www.pcworld.com May 2006, 45.

7 Molecular Economy

The world has seen massive development of computer and network technology, linking us to others throughout the world. Computer software is becoming more complex, capable of acting much like the computer is thinking on its own. Autonomous software is linked globally. This enables managers to shift from the traditional role of measuring and trying to predict, to a more liberated world where they can monitor events in real time and have the ability to respond in real time. This shift is referred to as adaptive management.[1] We discussed this general trend in Chapters 5 and 6. A second major movement is found in the development of what Meyer and Davis call molecular technology, by which they mean biotechnology, nanotechnology, and materials science. Meyer and Davis expected the convergence of these two developments to lead to a creative destruction in the vein of Joseph Schumpeter, leading us to a brave new age where things will be very different, forcing most of us to change what we do, but also providing us opportunities to do more things, much better.

The molecular economy applies microtechnology to business applications. This is the highest level of convergence known today, involving the convergence of biological systems and artificial systems to improve human lives.

Molecular Technology

Biotechnology has provided many beneficial tools in a variety of areas. It has led to scientific study in the fields of medicine, agriculture, and bioremediation, in addition to other fields. In medicine, organisms can produce antibiotics and genetic cures through manipulation of genomes. This includes pharmacogenomics which is the developing of drugs through genomic science.[2] A related application is gene therapy which is the insertion of functional genes into cells and tissues to treat disease.[3] In agriculture, it has led to transgenetic plants that have the potential to produce more environmentally friendly food crops. Genetic modification of rice was accomplished in 1982, with the development of golden rice which overproduces beta-carotene, useful in treating Vitamin A deficiency which can cause blindness. This has led to a major world controversy, as Europeans are quite averse to genetically modified foodstuffs, and refuse to allow any imports that might possibly be genetically modified. African nations that could benefit from golden rice have rejected it, to assure that they do not sour European trade relationships. Greenpeace has assumed that golden rice is a ploy to gain wider acceptance of genetic modification.[4] Biotechnology has also been used to aid bioremediation, such as having enzymes work on oil spills. Biotechnology also can be put to use for industrial purposes, to produce chemicals or to destroy hazardous chemicals.[5]

Nanotechnology is an applied field of science interested in controlling matter at the atomic and molecular scale. It was defined by its founder as "the processing of separation, consolidation, and deformation of materials by one atom or by one molecule."[6] Technological application was furthered in the 1980s and continues.[7]

Materials science is an interdisciplinary field of study of the properties of matter.[8] In the past, studies focused on metallurgy. However, with the computer age, interest has broadened to ceramics, polymers, magnetics, and all classes of materials. This has led to developments in semiconductor manufacture, medical implants, and biological materials.

Extensive progress in nanotechnology, biological engineering, and robotics has been seen as key to new military developments as well. While it would be nice to see a world without war, it must be recognized that military research has led to many breakthrough scientific gains, and in fact is a major source of research funding. Improved computers, sensors, and communications systems are key to military research advances.[9] Nanotechnology and biotechnology have led to chemically assembled electronic nanocomputers, potentially billions of times more powerful than current computers.[10] We also are all aware of advances in materials technology and its role in stealth aircraft.

Synthesis

Ray Kurzweil proposed that three overlapping revolutions were leading society towards what he called the "Singularity", where mankind would transcend the constraints of disease and hunger.[11] These revolutions involved genetics, nanotechnology, and robotics. The only difference between Kurzweil's vision and that of Meyer and Davis is Kurzweil's use of robotics as opposed to materials science. The difference in this last element is minor. Kurzweil as a computer scientist is more focused on the application of computers. Certainly great gains have been obtained in both computer applications to operate machinery and gains of the materials they work on.

Kurzweil defined combination as the essence of genetics. DNA is nature's nanoengineered computer. The new field of molecular computing was exemplified by mathematician Leonard Adelman, who managed to get a test tube of actual DNA to solve the traveling salesman (shortest-route) problem.[12] The shortest-route problem determines the shortest route from an originating point through a network to a destination, given the distance of branches in the network. A travel plan from an originating city to a destination city over various highways and through connecting cities, with distances between each city given, is a typical shortest-route problem. The objective is to find the route which will minimize total travel distance or time.[13] This problem can take a significant amount of time to solve for large numbers of sites. It is a combinatorial problem, which is the type of problem genetic algorithms can be effective in solving. Genetic algorithms were inspired by biological natural selection. Adelman and his research group seem to have inverted the logical stream of progression to utilize biology to apply genetic algorithms directly, through identifying enzyme reactions corresponding to arithmetic and logical operations. Molecular computing is being applied to additional useful problems.

Adelman's use of biological tissue to accomplish genetic algorithm solution to problems was done on a computer by John Holland, a key player in the founding of the Santa Fe Institute. Holland, a University of Michigan professor, was able to apply the ideas of viewing problems as genetic, with mutation of genetic alleles a way to consider new solutions, whose value was measured by some function, used to determine natural selection.[14] The method of Adelman and Holland is identical—the means of applying it is different. Other work at the Santa Fe Institute has been applied to economics, to include that of Stuart Kauffman.[15] Kauffman takes a systems view of the economy, consisting of autonomous agents coevolving into a web of diverse complexity to form the economy. Conventional economic theories of general equilibrium are insufficient to deal with this dynamic web. Kauffman suggests complexity theory as a way to model this adaptive economic system.

Molecular Economy

Molecular technologies are being exploited by a number of businesses. These advances are found through miniaturization, visualization, manipulation, and evaluation. Miniaturization leads to application in a number of areas. We are all familiar with the advances in computer chip technology, and Moore's Law, characterizing the rate of improvement in so many different aspects of computer technology. Miniaturization also has impacts in health care, through gene splicing and molecular biology.

Visualization advances come about from molecular technology as well. X-ray crystallography enables resolution to about the size of a water molecule.[16] Atomic force microscopes allow scanning surfaces to detect variances in atoms, which are then translated to an image. Commercial applications include tools to understand differences in chemical forces on surfaces, which can be useful in development of drugs or materials.[17]

Manipulation extends the knowledge gained through visualization. The NanoManipulator combines virtual reality and tactile feedback, allowing operators to manipulate material on a nano scale. On a larger scale, it is used in medical training to allow students to experience the feel of inserting needles into spinal columns.

Evaluation is yet another next step of gaining value from molecular technology. The genome project is one development coming from molecular technology, as are microarrays, hybrids of biotechnology and IT. A test medium is divided into thousands of partitions, with a different molecule attached to each section, binding to one of the targets. A chip gathers feedback from the probing molecules and generates a digital output. This ability for remote sensing is usable in detecting the presence of impurities, or when mechanical parts are wearing out, or when other specification limits are exceeded. In the medical field, molecular translation of measures into simulations of human organs allows experimentation in virtual space rather than real bodies.

Evidence of Convergence

The molecular economy is concerned with bringing business applications to microtechnology. Science has made many advances at the nanotechnology scale. Biotechnology and nanotechnology refer to areas of scientific advance. These advances have been especially effective in materials science and in automated computer applications such as robotics.

Ray Kurzweil saw these developments from the robotics side. He anticipated a time (he predicted sometime in the 2040 decade) when computers would transcend humans in the ability to think. Rather than being frightened of such a state, he viewed it as a golden era, when humans could allow computers to do much of the difficult decision making in life, allowing humans to greater levels of self-realization.

Meyer and Davis saw these developments leading to an age when there would be a convergence of information, biology, and business, leading us to evolutionary engineering. They saw the dynamics of self organization, recombination, selective pressure, coevolution and emergence leading to an entirely different world. The ability to make robots has been with us for some time. The new molecular economy would involve technology evolution beyond our current comprehension.

References

1 C. Meyer and S. Davis. *It's Alive: The Coming Convergence of Information, Biology, and Business*. New York: Crown Business, 2003.

2 R. Rapley and S. Harbron. (eds) *Molecular Analysis and Genome Discovery*. Chichester: John Wiley, 2004.

3 M.I. Phillips. (ed.) *Gene Therapy Methods*. San Diego: Academic Press, 2002.

4 E. Baggott. "A wealth deferred," *Harvard International Review* 28, No. 3, 2006, 28–30.

5 R.L. Oliver. *The Business of Biotech and How to Profit From It*. New York: McGraw-Hill, 2003.

6 N. Taniguchi. "On the Basic Concept of 'Nano-Technology'," *Proc. Intl. Conf. Prod. Eng. Tokyo*, Part II, Japan Society of Precision Engineering, 1974.

7 K.E. Drexler. *Nanosystems: Molecular Machinery, Manufacturing, and Computation*. Chichester: John Wiley & Sons, 1992; L. Gasman. *Nanotechnology Applications and Markets*. Boston: Artech House, 2006.

8 R. Asthana, A. Kumar and N.B. Dahotre. *Materials Processing and Manufacturing Science*. Boston: Elsevier Academic Press, 2006.

9 M.D. Mandeles. *The Future of War: Organizations as Weapons.* Washington DC: Potomac Books, 2005.

10 K. Chang. "IBM. Creates a Tiny Circuit out of Carbon," *New York Times* 27 August 2001.

11 R. Kurzweil. *The Singularity is Near: When Humans Transcend Biology.* New York: Penguin Books, 2005.

12 R. Kurzweil. *The Age of Spiritual Machines: When Computers Exceed Human Intelligence.* New York: Penguin Books, 1999.

13 S.M. Lee and D.L. Olson. *Introduction to Management Science.* Cincinnati: Southwest, 2006, 282.

14 J.H. Holland. *Adaptation in Natural and Artificial Systems.* Cambridge MA: MIT Press, 1992; J.H. Holland. *Hidden Order: How Adaptation Builds Complexity.* Cambridge MA: Perseus Books, 1995.

15 S. Kauffman. *Investigations.* New York: Oxford University Press, 2000.

16 M. Gross. *Travels to the Nanoworld: Miniature Machinery in Nature and Technology.* Cambridge MA: Perseus Books, 1999.

17 C. Meyer and S. Davis. *ibid.*

8 Information Technology Supporting Convergence

The premise of this book is that the functionality of technology (including IT) has led to many new opportunities to do things better within organizations. This expands the scope of management, presenting a need to think globally. The evolution of management can be viewed as:

1. management of the spot: optimization at the work station, task, factory, and functional levels;

2. management of the line: adoption of best processes;

3. management of the plane: coordination at the inter-organizational level;

4. management of space: coordination of global value chain networks.

Success calls for the convergence of technological developments with innovation. This convergence leads to a flatter world, with greater opportunities for organizations and people to advance economically.

Value chains have been effective in reorganizing how products are delivered to customers. Rather than the 19th century focus on vertical integration, providing control through ownership, contemporary relationships are obtained through strategic alliances, as world-class competencies can be pooled to create cooperating organizations, each doing its part in coordinated networks of loosely linked organizations. Such value chains could be in the form of joint ventures, application service provider relationships, service oriented architecture, and so on. Some of the most effective retailing organizations in the world are of this type. Some are more centrally controlled, such as Wal-Mart. Others are more loosely coordinated, such as Nike and Mattel. Virtual communication of information and data make it possible for these networks to coordinate and innovate. Global value chain innovation can include a variety of forms at different organizational levels as we discussed in Chapter 4:

1. Adaptation: maximization of local relevance.

2. Aggregation: globally consistent organizational strategy formulation.

3. Arbitrage: leverage regional differences and local competitive advantages.

Procter & Gamble is an example of an organization leveraging local competitive advantage of its supply chain alliances into a cohesive, efficient global strategy.[1] Procter & Gamble views its core capability as product development and commercialization. Manufacturing is not considered a core capability, and therefore they outsource most manufacturing. Nike operates in a similar manner, employing few people directly, yet orchestrating complex global value chains to manage their brand of products. An example of Nike alliances is their collaboration with Apple Computer, offering a Nike running shoe named the Moire, which includes a sensor connected to a small wireless receiver attached to Apple's iPod®.[2] The system can tell runners how far they've gone, and at what pace, over the iPod® system while delivering the runner's favorite music. Nike also has a website where runners can track their progress, set goals, and share their results. Nike's philosophy is an example of convergence used to generate innovative strategy, focusing on "Can we harness the power of digital technology to improve a runner's experience?"[3]

This chapter will discuss how the new world order involves technological equality, as business entities around the globe can join with other organizations in cooperative efforts. This is a crucial element in convergence within supply chains, which have seen many uses of IT to improve management of goods/services over complex supply chains. The link to strategic innovation within supply chains is reviewed. Potential for expert system-related products is then presented.

Technological Equality

The industrial revolution did not spontaneously erupt around the world. Competitive advantage in technology and capital in the 19th century was best developed in England, spreading throughout Europe and North America. This is not to say this was the original development of industry. China had dominated advanced technology for many centuries, but was later isolated by distance and policy from the rest of the world.[4] There were many well-known developments in the 20th century, leading to the rise of Western Europe, then the

United States and Japan in dominating technology-based production of goods and services.

In the last two decades of the 20th century, there was a notable shift in production to south-east Asia as new technologies became available throughout the world, and transportation networks enabled practical global ability to transfer materiel. Technologies in the form of equipment (tractors, combines, and so on) as well as virtual technology (Microsoft Office, the internet and all of its related IT capabilities) emerged. This enabled convergence of the type described in Chapter 3: the combination of technology with new methods of producing goods and delivering services, such as the cell phone. The competitive advantage used to be largely dependent on technological advantage. Technological equality has brought a relatively level field of competition for organizations around the globe. Thus, convergence of what is known and unknown has become a new source of competitive advantage.

The cell phone represents this convergence through its ability to take pictures (and send them instantaneously around the world), to link to the internet to check e-mail, to not only view product/service availability (such as books or restaurant reservations), but to take action (such as making a purchase). The potential exists for businessmen to expand what they do with their Blackberry® to access data files, giving them the potential to virtually work from the beaches of Tahiti. Technological superiority no longer provides sufficient competitive edge for organizations. It is necessary to innovatively combine and leverage technology with the way in which business is done. The future belongs to those who can come up with new knowledge and better ways of doing things. And the rewards of such innovation are going to be short-term, leading to the need to come up with better ways of doing things on a continuous basis. The iPhone® is a case in point. Apple brought out their new technology tied to AT&T service in early 2007. In the second half of 2007 a hacker figured out a way to access service bypassing AT&T. Undoubtedly Apple and AT&T will try to make this bypass *passé*. It is a global technology arms race. In the process, hopefully the consumers of the world will benefit from this motivation to improved methods.

In the age of technological equality, the real competitive advantage will come not from using the technology to do the conventional business functions, but rather from creating and selling new knowledge and ideas. Information technology opens new doors to possibilities of accomplishing things never dreamt of before. This chapter reviews some convergences of function with technology in the information system area. These include enterprise information systems (EIS), which provided value to organizations on a massive scale, making vendors like SAP and Oracle members of the world's largest volume businesses in terms of sales. These systems were widely applied to large organizations in the 1990s, with a focus on large dollar tag customers. Since then, vendors have emphasized web access, a form of the convergence we are talking about. Supply chain operations provide value to manufacturing as well as retailing, enabling access to more economical sources and partners around the world. By their nature, supply chains involve convergence of a process that has existed for centuries with electronic technology and other innovations. We will review a sequence of convergence techniques that allow greater efficiency in supply chains.

Supply Chain Convergence

E-business has brought us an entirely new organizational climate, with the ability to electronically generate retail data. The convergence of IT with supply chain operations in retail led to opportunities for Wal-Mart to gain a competitive advantage, which enabled them to transform the retail environment throughout the United States.[5] Many service organizations and retail outlets generate masses of data. Grocery stores have large amounts of data generated by our purchases. Bar coding has made checkout very convenient. Wal-Mart and other retailers have extended electronic data generation to include radio frequency identification (RFID), allowing tracking of product physical location in real time.

Electronic data can be—and are—shared across elements of supply chains. This provides the potential for more complete system optimization. The requirements for successful data sharing, however, constrain this potential. First, members of the supply chain need to have compatible information systems. Since the core members of a supply chain may have very sophisticated software, this may force out smaller participants who cannot afford the same level of software sophistication. There also is added complexity in the system, and uncertainty in demands may lead to unstable systems. Seeking to optimize systems with highly uncertain data is problematic.

There are about five representative forms of supply chain coordination in the electronic business environment. These systems certainly apply to the service sector in the broad sense. Successful retail organizations in the 21st century consider service to the customer to be their focal point. Electronic management across supply chains enable them to provide better service.

Traditional Supply Chains

There have been a variety of control schemas that have evolved to control supply networks. The traditional uncoordinated supply chains of the 1980s, with no information sharing, but independent inventory control policies, led to the infamous "bullwhip" phenomenon.[6] The bullwhip phenomenon occurs because of overestimation of demand induced by the lumpiness of orders from downstream elements of the supply chain. An obvious first step in reducing the inefficiencies due to the bullwhip effect was to increase information sharing across the supply chain.[7]

Efficient Consumer Response

The first type of coordination among supply chain members is information exchange, to include action plans to enable forecast alignment for long-term and capacity planning.[8] This improves visibility and thus makes demand more predictable. Efficient consumer response (ECR) transformed business practices in the grocery industry through implementation of four strategies:[9]

1. efficient promotions;
2. efficient replenishment;
3. efficient store assortment;
4. efficient product introductions.

In some cases inventories have been reported to have actually increased through the use of ECR.[10] One source of distortion can arise when retailers give false data to distributors or manufacturers in efforts to assure more secure supplies.

Vendor Managed Inventory

Vendor managed inventory (VMI) involves supplier assumption of management of retailer inventory.[11] This is channel coordination. Based upon advanced information through electronic data interchange or the internet, the supplier controls the stock at the retailer. VMI can lead to shipment consolidation.[12] Both retail customer service levels and inventory turnover can be improved through VMI, and through more accurate forecasting and more effective distribution of inventory in the supply chain.[13] VMI would allow retailers to expand the assortment of products they carried in a given retail space, thus improving brand profitability for both retailer and vendor.

VMI was adopted by many firms, including Procter & Gamble and Wal-Mart.[14] However, problems have been reported in some firms and VMI has been abandoned by them. VMI works well when manufacturers supply large volumes of frequently replenished product under relatively stable sales conditions. But high levels of demand volatility can lead to excessive inventories, the same problem that existed in traditional retail inventory control through the bullwhip effect.

Continuous Replenishment

Continuous replenishment (CR) was proposed to implement automatic replenishment programs, where sellers restocked retailer inventory based upon actual product usage and stock level information provided by the retailer. Wal-Mart piloted CR in 1995 and it has been used by many US and UK retailers.[15]

Retailers make point-of-sales data available to suppliers, allowing inventory decisions based on sales forecasts rather than on inventory level variations. CR enhanced VMI by requiring supply chain members to share more information and data, and to utilize common systems and use common performance measures. This promoted joint decision-making, accountability, and incentives for performance. CR has been cited as successful in improving customer service levels and inventory turnover.[16] However, because CR does not necessarily cover inventories throughout the supply chain, it still can include gaps. Manufacturer prediction of future retail events is the major missing feature of CR. Excess inventory seems to be shifted from retailers and distributors to manufacturers. While CR improved VMI, further benefits were available.

Collaborative Planning, Forecasting and Replenishment

Collaborative planning, forecasting and replenishment (CPFR) was the next step, applied in the drug, grocery, apparel, and other industries.[17] The manufacturer and retailer exchange marketplace information to develop customer-specific plans to substantially reduce inventory. Promotion schedules, point-of-sale data, and inventory data are shared to enable shortening lead-times and integration of forecasting and replenishment. Thus total visibility is obtained, and changing demand patterns can be considered. Warner-Lambert, Wal-Mart, SAP, Manugistics, and Benchmarking Partners initiated the first CPFR pilot.[18]

The way in which information is exchanged earlier in the process yields a number of benefits.[19] For retailers, these include increased sales, faster response times to orders, higher service levels despite lower inventories which in turn leads to lower costs from obsolescence and deterioration. Manufacturers benefit through increased sales, higher fill rates for orders, faster cycle times, lower capacity requirements, and lower product inventories. There are fewer stocking points needed over the supply chain, improved forecast accuracy, and overall, lower expenses over the system.

Boeing and Alcoa jointly implemented CPFR.[20] Alcoa supplied aluminum products for Boeing airplanes, using rail transportation which involved unreliable delivery dates. Close cooperation led to an atmosphere with high levels of trust. Boeing sent weekly electronic forecasts and inventory counts to Alcoa, and used its EIS system to generate electronic purchase orders for raw materials. Alcoa implemented a VMI system, and improved forecast visibility. The electronic forecasting system was credited with expediting information exchange. Boeing carefully checked its forecasting, and identified modifications to data obtained from their EIS that were needed to provide Alcoa with more accurate data. The greater degree of forecasting accuracy obtained enabled Alcoa to make more efficient production decisions.

The cooperation between Boeing and Alcoa demonstrates the potential of CPFR. As with VMI, systems need to be compatible across organizations, and the more organizations that are involved, the more restrictive this requirement is. However, CPFR can lead to vast efficiency improvements in participating organizations.

Convergence and Supply Chains

This evolution of supply chain techniques represents convergence on a large scale. Communication technologies in various forms rely on the internet to share data across members of supply chains. A sequence of techniques (shown in Table 8.1) emerged to answer specific problems. Not every supply chain in retail uses all of these techniques, but variants provide value to many of them.

Table 8.1 Supply chain coordination systems

System	Description	Applications	Problems
Traditional	Independent inventories, communication by purchase order	All types of business	Bullwhip
ECR	Independent inventory management, sharing information electronically	Grocery	Distorted data can be counterproductive, fast orders but slow replenishment
VMI	Vendor-managed inventory downstream (channel coordination)	Procter & Gamble, Wal-Mart Groceries	Need good data visibility, predictability, compatible systems, and trust. Doesn't work as well if substitutes readily available
CR	Automatic replenishment based on real-time data forecasting	Wal-Mart improved customer service levels with better inventory turnover	Manufacturer can't predict special retail events, excess inventory shifted from retail to upstream
CPFR	More complete data sharing, enabling vendor forecasting of downstream demand, joint planning	Drug, grocery, apparel Warner-Lambert, Wal-Mart, SAP, Manugistics Nabisco/ Wegman Foods Boeing/Alcoa	Need compatible systems Difficult to keep up in dynamic market-driven demand

These systems represent convergence, the use of evolving technology to conduct business in better, more effective ways.

Strategic Innovation in Supply Chains

Supply chain linkages have been used strategically in the past. The most dramatic success is by Wal-Mart, but many other examples exist, such as Zara in fashion clothing retail and by Li and Fung in trading.[21] Zara took a radically different approach in stocking inventory, which could be very expensive in the clothing business. Instead of stocking anticipated demand, they stocked minimal quantities of each piece of clothing and waited to see where the demand was. They developed the capability to quickly ramp up to meet demand for popular items, enabling them to avoid many of the risks of overstocking expensive items. Li and Fung developed similar agility in international goods, relying on internet data exchange and communication to operate more efficiently than their competitors.

The concept of supply chains is well-developed and the internet provides communications access to link organizations world-wide on a regular basis. The strategic innovation part comes about when people figure out new ways to utilize these opportunities to gain business advantage. This can evolve into a demand chain, when organizations such as Amazon.com use IT to nudge their customers to look at new products that are likely to be attractive to them individually. We expect that the brainpower of the world will continue to find innovative ways to utilize supply chain linkages.

Evidence of Convergence

Supply chain technology enables organizations to cooperate to provide goods and/or services to their customers. Convergence occurs because of the need for shared data and coordination. This often is accomplished through shared software systems, as is accomplished by Wal-Mart (who require participating vendors to have compatible software technology).

Information technology has evolved to provide many tools that help people better control organizations and to conduct business in a better manner. Computer systems come in many forms, and have been useful in enterprise systems, control of materials, and information flows within supply chains, aid to decision making in both unstructured and structured problem domains, and in recognition of new patterns through data mining. These technologies exist to help decision makers. The point of our book is that the convergences we brought up in Chapter 3 have been enabled by developments in IT. The future belongs to those who learn new technologies and spot opportunities to utilize them in ways to do things better.

A major venue of opportunity for convergence is provided by Web 2.0, with an emphasis on interaction of many-to-many as opposed to one-to-one interaction typical in the 20th century. A massive number of new environments for interaction exist, including Facebook and MySpace. New services are provided such as eHarmony.com, which require interaction of multiple parties to work. This represents a shift from a closed view of the world, concerned about privacy and security, to an open view. This open view has been present for some time, manifested in software products such as Linux, the widely known open source operating system. Open software systems have become widespread, with even ERP systems available for free downloading (Compiere). As newer computer users grow up in this environment emphasizing openness, we expect the trend to steamroll on and lead to ever greater opportunities for collaboration over Web 2.0.

References

1 "Partners in wealth," *The Economist* 378, No. 8461, 1/12/2006, 16–17.

2 A. Kesseldahl. "Apple and Nike, running mates," *Business Week Online* 5/9/2006, 9.

3 *Ibid.*

4 G. Menzies. *1421: The Year China Discovered America*. New York: Harper Perennial, 2004—discusses the Chinese policy change to cease global exploration after successful fleet operations.

5 C. Fishman. *The Wal-Mart Effect: How the World's Most Powerful Company Really Works – And How It's Transforming the American Economy*. New York: Penguin, 2006.

6 J.D. Sterman. "Modeling managerial behavior: Misperceptions of feedback in a dynamic decision making experiment," *Management Science* 35, No. 3, 1989, 321–39; D. Taylor, "Measurement and analysis of demand amplification across the supply chain," *The International Journal of Logistics Management* 10, No. 2, 1999, 55–70.

7 H.L. Lee, K.C. So and C.S. Tang. "The value of information sharing in a two-level supply chain," *Management Science* 46, No. 5, 2000, 626–43.

8 M. Holweg, S. Disney, J. Holström, J. and J. Småros. "Supply chain collaboration: Making sense of the strategy continuum," *European Management Journal* 23, No. 2, 2005, 170–81.

9 M. Barratt. "Positioning the role of collaborative planning in grocery supply chains," *The International Journal of Logistics Management* 14, No. 2, 2003, 53–66.

10 T.A. Brown and D.M. Bukovinsky. "ECR and grocery retailing: An exploratory financial statement analysis," *Journal of Business Logistics* 22, No. 2, 2001, 77–90.; R. Frankel, T.J. Goldsby and J.M. Whipple, "Grocery industry collaboration in the wake of ECR," *The International Journal of Logistics Management* 13, No. 1, 2002, 57–72.

11 T. Andel. "Manage inventory, own information," *Transportation and Distribution* 37, 1996, 55–8.

12 S. Cetinkaya and C.Y. Lee. "Stock replenishment and shipment scheduling for vendor-managed inventory systems," *Management Science* 46, No. 2, 2000, 217–32.

13 D.D. Achabal, S.H. McIntyre, S.A. Smith and K. Kalyanam. "A decision support system for vendor managed inventory," *Journal of Retailing* 76, No. 4, 2000, 430–54.

14 J.A. Cooke. "VMI: Very mixed impact?" *Logistics Management and Distribution Report* 37, No. 12, 1998, 51.

15 M. Barratt. *ibid.*

16 R.C. Vergin. "An examination of inventory turnover in the Fortune 500 industrial companies," *Production and Inventory Management Journal* 39, 1998, 51–6.

17 R. Ireland and R. Bruce. "CPFR: Only the beginning of collaboration," *Supply Chain Management Review* 4, No. 4, 2000, 80–8.

18 M. Barratt. *ibid.*

19 G. Fliedner. "CPFR: An emerging supply chain tool," *Industrial Management & Data Systems* 103, No. 1, 2003, 14–21.

20 V.A. Micheau. "How Boeing and Alcoa implemented a successful vendor managed inventory program," *The Journal of Business Forecasting* Spring 2005, 17–19.

21 J. Gattorna. "Supply chains are the business," *Supply Chain Management Review* 10, No. 6, 2006, 42–9.

9 Innovation through Open Systems

Open Internet Systems

The advent of Web 2.0 opened another level of openness leading to collaborative participation. Web 2.0 includes three key properties: openness; participation; and collaboration. The meaning of openness defined in a Web 2.0 context is broader than in the pre-Web 2.0 era, which included mainly open-source capabilities, where anyone with access to the web could obtain software. Openness in the Web 2.0 era—in addition to open-source—includes open innovation, crowdsourcing, and so on. The traditional path to innovation was the accumulation of knowledge developed in the R&D department, which was secured inside the organization to secure market competitiveness. Open innovation is a new phenomenon that companies need to be successful in the new Web 2.0 environment, relying on interactive participation by multiple users to generate new and better solutions. Crowdsourcing is inviting the public to take on the tasks that were traditionally performed by employees or contractors. The perceived benefits of crowdsourcing (Wikipedia) are a comparatively low cost to explore given problems, payment by results, and the ability to tap into a wide range of talent.

In Web 2.0, information consumers are becoming information providers. Prosumers (producer and consumer or professional and consumer), highly skilled hobbyists before Web 2.0, are being replaced by Pro–Ams (between professional and amateur) who are creating contents of value that contribute to society and the economy. The human brain has limited capacity to process the abundance of information generated. The massive amount of information available from Web 2.0 diverts attention. Attention itself is the most important value in the Web 2.0 world. How to create new business models based on the new value is indeed a critical question.

Once openness is adopted as a business strategy then participation and collaboration naturally follow. Blogs, mashups, and wikis are the three important types of Web 2.0 collaborative tools.[1] Blogs frequently provide a combination of a person's personal life and the particular subject they are currently interested in. Blogs have become a powerful tool for establishing and maintaining online communities. Their audience ranges from family and friends to professionals, departments and the general public.[2] Mashups can be defined as "creating new applications by combining different resources online."[3] Many mashups are created using Web applications that have a publicly available Application Programming Interface (API). By writing a middleware program, information from two or more web applications can be combined. For example, Google provides APIs for many of its services (http://code.google.com/). As one example, there have been several mashups that have combined Google Maps and publically available crime data to automatically generate maps showing the locations of crime in certain cities. Wikis are another type of collaborative software applications that build websites. A wiki enables a group to collaboratively develop a website, such as a project site, knowledge base, or resource. Wikipedia is one of the most familiar wiki sites. Wikinomics consists of four powerful new ideas: openness, peering, sharing, and acting globally.[4] These all enable mass collaboration over the web.

Prior to Web 2.0, computer mediated communication (CMC) included advertising, and blogging, and so on. These were based on the typical industrial economy system that was closed, uniform, and had a one-way value chain. The invention of the internet and Netscape opened new pathways for businesses to reach online markets. However, the collapse of countless dot-com companies resulted from inflexible, one-way communication. Instead of seeking economies of scale, businesses in the Web 2.0 environment should focus on the new measure of economic value, attention. The development of digital equipment and the ease of using the web led to an increasing number of content users becoming content providers. With mass media, content providers post their material to the open web, allowing content consumers to read their material (usually with the help of a search engine). This is a one-way flow of information. Mass media is not the only means to broadcast. The web is full of user-generated content reaching everywhere. Podcasting is also taking the place of broadcasting. Users are participating in events around Web 2.0 in venues such as social networking sites, user generated content, and so on.

Service Oriented Architecture

Service oriented architecture (SOA) is a design approach for technology convergence. OASIS (Organization for the Advancement of Structured Information Standards) defines SOA as: A paradigm for organizing and utilizing distributed capabilities that may be under the control of different ownership domains. It provides a uniform means to offer, discover, interact with and use capabilities to produce desired effects consistent with measurable preconditions and expectations. SOA unifies business processes by structuring large applications as an *ad hoc* collection of smaller modules called services. These applications can be used by different groups of people both inside and outside the organization, and new applications built from a mix of services from the global pool offer greater flexibility and uniformity. This enables many users to apply the same computer routines for online checking, savings account transactions, completing online applications, viewing statements, placing orders, and so on. The computer system architecture is designed so that each functional element is not dependent on any of the others. Computer system designers call this "loosely coupled."[5] It provides a standard architecture so that computer system components built by others who comply with these standards will function in a system. An example of how this can lead to the convergence of technology to better ways of conducting business is the telecom domain.[6] New technology tools enable new ways to collaborate, customize services, and personalize products. Service development life-cycles are being reduced, and autonomous service assets are capable of being composed into flexible offerings to customers. This enables imaginative users to transform their business continuously.

Web Services provide a convenient way to access existing internal and external information resources. It uses a number of technologies[7] to build programming solutions for specific messaging and application integration problems. However, building a new information system is in some ways like building a new house. Web Services is analogous to cement and bricks. Blueprint and engineering knowledge are more important. SOA gives the picture of what can be done with

Web Services. SOA exploits the business potential of Web Services. This can lead to a type of convergence by enabling using organizations to access better methods at lower cost through technology.

SOA is a strategy based on turning applications and information sources which reside in different organizations, systems, and execution environments into "services" that can be accessed with a common interface regardless of the location or technical makeup of the function or piece of data. This interface must be agreed upon within the environment of systems that can access or invoke that service. A service within SOA either provides information or facilitates a change to business data from one valid and consistent state to another one. Services are invoked through defined communication protocols. The pivotal part of SOA is how communication between different data formats can be accomplished. Web Services, which is independent of operational environment, allow this communication.

To compete, businesses need to react to changes quickly. For example, in a value chain, different business partners should be dynamically added or removed as required.[8] Volatile business needs aggravate "impedance mismatch" or gaps between different business partners" IT systems. To respond with speed under conditions of constant change, existing business processes as well as IT systems quite often need to be updated and even replaced. However, constant innovation with traditional IT systems is almost impossible in practice. Traditional computing models focus on highly structured tasks and are able to serve rigid environments very well. They will break down in unstructured environments.

IBM calls this new business model On Demand Business. Hewlett-Packard calls it Adaptive Enterprise. Dell does something a bit more focused on expansion that it calls blade computing. The essence is that firms do not need to maintain huge, complex and expensive computer systems whose capacity sits idle most of the time and cannot communicate with the computers of suppliers and customers. Computing needs for business applications and corporate data centers would be served by drawing on computational resources supplied by a combination of specialist service providers, enterprise resources, and small-scale opportunistic suppliers. With this on-demand model, a service is "brought-on-line" if and when is needed and solely for the duration of its need.

SOA Leading to New Products

Innovative technologies such as SOA support re-engineering of information infrastructure. Bielski referred to this as generating the positive aspects of growth in the technology sector, along with convergence in the consumer market with products such as the iPod®.[9] SOA provides the opportunity to improve functionality of applications, using information technology transformation for applications such as financial services. This can include applications such as fraud management, made more effective through analysis of large scale databases of claims through data mining.

In the banking industry, web businesses have appeared, such as Salesforce, which has over 1,500 financial services customers. Examples of their customers include Commerce Bank, which wanted to automate forecasting processes related to customer relationship management. Their criteria included assurance of an open but secure system that would not compromise sensitive financial data. Through SOA, Salesforce was able to create a computer application satisfying these requirements without any additional information technology requirements on the part of Commerce Bank. SOA allowed access to particular expertise accessible without problems of compatibility.[10]

Another example is Bank of the West, which had a manual human resources procedure that they felt was excessively expensive. By changing the way in which they did their work, through computer software accessible because of SOA, they made improvements in their operations. SOA enabled access of software enabling better identification of seasonal traffic variation and other factors. By accessing software developed for specific functions, Bank of the West was able to save US$8 million over two quarters through more accurate information and reduced administrative costs.

SOA can also make access of radio-frequency identification (RFID) devices to better track inventory.[11] RFID technology has been widely applied by organizations such as Wal-Mart to change the ways in which inventory is tracked and managed. In order to work, standard architectures are needed to communicate data across computer systems that are used across complex supply chains.

Methods Improvement from On-Demand Computing

On-demand business means that any qualified user can access any program any time from any place.[12] With such a business model, firms do not need to maintain huge, complex and expensive computer systems that often sit idle and cannot communicate with the computers of suppliers and customers. Computing needs for business applications and corporate data centers would be served by drawing on computational resources supplied by a combination of various service vendors. A service is "bought-on-line" if and when needed and solely for the duration of its need.[13] The computing model provides a mechanism to link various component services which are created and run in different operation environments. It creates a highly modular environment, within which application and infrastructure components can be more easily defined and managed.[14] Through the composition of components, a flexible and real-time implementation of business policies becomes more practical than in the past.

Web Services and SOA are dominant technologies supporting this new business model. Web Services is a collection of technologies that is supported by tools such as XML (Extensible Markup Language), SOAP (Simple Object Access Protocol), WSDL (Web Service Definition Language), UDDI (Universal Description, Discovery and Integration), and so on.[15] These technologies can be used to build programming solutions for specific messaging and application integration problems. SOA is an architectural paradigm based on turning applications and information sources which reside in different organizations, different systems and different execution environments into "services" that can be accessed with a common interface regardless of the location or technical makeup of the function or piece of data. Services can be a variety of things, to include quote services, communications protocols, and other computing functions.[16] Differences between these concepts and traditional systems are briefly outlined in Table 9.1. The primary issue of SOA implementation is how to realize communication among different data formats. Web Services solves this problem. It is independent of operational environments.

Table 9.1 Comparison of traditional and on demand systems

Traditional Systems	**On-demand Systems**
Self-contained, comprehensive software	Web Services XML SOAP WSDL UDDI
Organizations obtain independent software products. More recently, ERP with add-ins	Service Oriented Architecture, quote services, communications protocols

On demand business is based on a new computing model which revolutionizes the way of communicating among different parties' systems. It provides a simple mechanism to link various components which are created and run in different operation environment. This enables the creation of a highly modular environment, within which application and infrastructure components can be more easily defined and managed.[17] At the same time, through the composition of components, a more flexible and real-time implementation of business policies becomes more realistic than before.

IBM gives three principles in implementing an on-demand model.

1. To achieve more flexibility and innovation, organizations need to become more componentized. The overall business should be broken down into its constituent components. Users no longer look at their business through lenses like organization, geography, and product or customer

segment, but begin to look at it through the lens of discrete functions and processes.

2. Applications should become more modular. Instead of focusing on big monolithic applications made up of rigid functional parts, users begin to think about discrete modular elements of application functionality that can more easily be modified and composed at the real time pace needed by businesses.

3. To support componentized business models and modular applications, an operating environment is required that can speed the integration process, simplify overall complexity and reduce the unnecessary consumption of resources.

SOA and EIS

Web Services is a technology supporting new business models such as On-Demand Business and Adaptive Enterprise. Its primary advantage is to fully make use of various IT resources through creating bundles of different component services. Even ERP support is available from open sources through web access.[18]

Enterprise Information Systems

Among the most complex computer systems supporting organizations are enterprise information systems (EIS, often referred to Enterprise Resource Planning, or ERP). These systems integrate organizational computing by a unified architecture of servers to share information and common databases that lead to more accurate information securely shared throughout the organization.

There are many reasons to adopt an EIS. EIS offers an integrated system shared by all users rather than a diverse set of computer applications, which rarely can communicate with each other, and with each having its own set of data and files. EIS provides a means to coordinate information system assets and information flows across the organization. The main benefit is the elimination of sub-organizational silos that focus on their own interests and problems rather than serving the interests of the whole organization. On the downside, EIS impose one procedure for the entire organization, which requires everyone to conform to the new system. EIS are thus less flexible. But the benefits of integration are usually much greater than the costs of conformity.

Data can be entered once, at the most accurate source, so that all users share the same data. This can be very beneficial, because shared data is used more often, by more people, which leads to much more complete and accurate information. As errors are encountered, users demand correction. There are limits, as a set of procedures are needed to see that changes do not introduce new errors. This makes it harder to make corrections, but again, this added inconvenience is usually well worth the gains of data integration.

EIS also can provide better ways of doing things. This idea is the essence of best practices, a key SAP system component. The downside to best practices is that they take a great deal of effort in identifying the best way to proceed with specific business functions, and that they often can involve significant change in how organizational members do their work. Further, as with any theory, what is considered best by one is often not considered best by all. What is "best" is also dynamic, as with time new and better ways of doing things (sometimes due to technological development) are identified. Organizations adopting EIS thus need to consider the permanence of best practices. Vendors may offer better "best practices" in their next version.

Enterprise systems are usually adopted with the expectation that they are going to yield lower computing costs in the long run. Ideally, adopting one common way of doing things is simpler and involves less effort to provide computing support to an organization. Improved methods through business-process re-engineering are an example of convergence to find new ways to do things better. But it is difficult to find the best practices for each organization. In practice, savings are often not realized, due to failure to anticipate all of the detailed nuances of user needs, as well as the inevitable changes in the business environment that call for different best practices and computer system relationships. Training needs are typically under-budgeted in EIS projects. Furthermore, these training budgets don't usually include the hidden costs of lost productivity as employees cope with complex new systems. Another problem area is that vendors are constantly seeking to improve their products, but that means that they will drop support on their older versions, forcing clients to face large unanticipated expenditures to maintain their systems. Table 9.2 recaps these pros and cons of EIS.

The key rationales for implementing EIS are the use of technology to provide more powerful, integrated computer systems. Current trends emphasize web access for users, a form of convergence of information technology with information systems that provides greater flexibility with likely improvements in cost performance. Concurrent with information system integration, business practice reengineering can supplement the opportunities to improve operational quality and/or greater productivity, another variant of convergence of function and technology.

Table 9.2 EIS pros and cons

Factor	Pro	Con
System integration	Improved understanding across users	Less flexibility
Data integration	Greater accuracy	Harder to make corrections
Best practices	More efficient methods	Imposition of how people do their work, less freedom and creativity
Cost of computing	More efficient system planned	Changing needs, under-budgeted training expense, maintenance upgrade expense, hidden costs of implementation

Strategic Innovation and EIS

Vendors of enterprise systems have saturated the market with their best practices. One view is that this saturation has reached the point that IT no longer matters, because now everyone has access to the same software tools.[19] This would be true if the vendors didn't place such high price-tickets on their products. There is a price opportunity for competitors to SAP and Oracle for many small-to-medium sized organizations that Microsoft and many other competitors are moving into.[20] But there is a valid point in that EIS has been sold to be a tool to take care of day-to-day operations at the lower levels of the organization. IT in that sense becomes commoditized. Strategic value is not obtained from doing things as well as everyone else is doing.

However, EIS can be used strategically. EIS is like a fine fabric. A creative use of the fabric to make high fashion clothes can take many different forms. Creating the best value from EIS represents a strategic use of IT. This is not automatic, but takes careful analysis and planning. Companies with the biggest IT investments rarely post the best financial results. But those that think smart rather than doing everything their SAP sales representative recommends can gain competitive advantage. Information technology offers great opportunities to do things better. This can be strategic if it is done in a more cost effective way than done by competitors.

EIS and Convergence

Enterprise information systems by their nature involve convergence on a macro scale. They are the implementation of information technology to provide information to organizations in a much better way. Prior to EIS, organizations had large information-systems staff responsible for systems analysis, programming, testing and other functions needed to develop their own software. This environment was improved a great deal with the development of COTS (commercial off-the-shelf software products) that were programmed by their vendors to provide specific functionality, such as banking software, payroll software, and so on. However, SAP, BAAN, and J.D. Edwards integrated the basic software functionality needed to run an organization. This involved massive research, especially by SAP, to find what they thought was the best way to do specific tasks (best practices), which each vendor programmed into their enterprise resource planning software. SAP spent considerable research efforts to identify the best way of doing conventional enterprise resource planning tasks. They have 800–1000 best practices included in their R/3 software.[21]

Consultants often develop further specialized expertise that firms can purchase. A best practice is a method that has been judged to be superior to other methods. This implies the most efficient way to perform a task.

On a more micro scale, each organization that adopted an EIS in the 1990s tended to apply business process re-engineering (BPR) to identify the best way to accomplish tasks in their specific organization. Again, this involved the implementation of IT to do things in a different way, yielding a much more efficient organization. BPR was developed with the idea of systematically identifying new and better ways to do things. In organizations, this usually involves the use of computers to replace old manual approaches centered on functions.

The focus of EIS has been on supporting operational functions. That is what vendors emphasize when they pitch their products to potential customers, because that is the easiest type of benefit to justify cost. However, the future of information/communications technology is through knowledge management and idea creation through convergence. Actually, there has been one case widely reported where an ERP system was adopted with the intent of strategically changing a firm's product, with the intent of gaining strategic dominance over their competitors. FoxMeyer Drug adopted an ERP in the mid-1990s to revolutionize how they provided wholesale drug products to their customers. The intent was excellent, and would have worked had they not botched the implementation project.[22] ERP systems can and are being used strategically, to give using organizations cost advantage through access to more complete information in a coherent, integrated information system. Those that are successful do not typically publicize that aspect of their success, preferring to emphasize their product's value.

Current Open-Source ERP

The goal of ERP is to integrate and consolidate all the old departments across an organization into a one system that can meet and serve each department's unique needs and tasks. Therefore, every aspect of an organization's business process will have a unified application interface, which provides high competitiveness in the market. Enterprises have invested heavily on ERP acquisition while small businesses or entrepreneurs often could not see the affordability of it mainly due to its high upfront prices and lack of resources to maintain the system. To attack this niche market of ERP in SMB sector, vendors has developed transformed ERPs by adopting the most advanced information technologies available. The most available business models of ERP include software as a service (SaaS), open source software (OSS) and service oriented architecture (SOA).

SaaS offers ERP as a service that clients can access via the internet. Smaller companies are spared the expense associated with software installation, maintenance and upgrades.[23] Mango Network, an Irving, Texas software and services company, is a channel of providing software and services for small and midsize wholesale and retail distributors. It combines the pure open-source business model and SaaS. Compiere which is a pure open-source company provides products and Mango sells them through SaaS. Mango charges annual fees based on a customer's revenue, rather than monthly fees based on the number of users.

OSS ERPs allowed small and medium sized businesses' access to ERP. The benefits of applying OSS are as follows:[24]

- Increased adaptability: Since ERP is not plug-and-play, implementation processes are necessary to match the company's business processes and local regulations. Having full access to the ERP source code is beneficial.
- Decreased reliance on a single supplier: Proprietary ERPs highly depend on the services from vendors and distributors. Upgrading and maintain service can be obtained from single source.
- Reduced costs: Proprietary ERP licenses are expensive. OSS ERPs' average implementation costs are at between one-sixth and one-third of the costs for typical proprietary ERPs.

Web 2.0 Business Models

The majority of Web 2.0 business models can be explained by focusing on the process used in Web 2.0, value creation production and filtering. Production produces value and filtering tailors information based on customers' needs. There were only a limited number of people, called professionals, who could participate in these steps in traditional industrial economic systems. Production was limited to reporters in the press or developers of software firms and filtering was limited to editorial writers, managers and marketers.

1. Limited Production model (free software, no participation by users)

    - all activities are done by the members/employees
    - guarantees high quality of final products and services after filtering.

2. Open model

    - everyone can be a producer
    - filtering can also be open to everyone
    - filtering process is more important because of the scarcity of attention
    - quality of final products or services is not guaranteed.

3. Open Production and limited filtering: hybrid collaboration model (such as Linux)

    - production is done by users and filtering is done by a small number of professionals
    - high quality products and services are guaranteed.

4. Limited Production and Open Filtering: Competition model

    - production is done by a small number of professionals or Pro–Ams and filtering is done by public users
    - filtering is pre-screened by professionals then open to public for voting
    - inducing public participation
    - public opinions are emphasized.

The most common business model of OSS is based on a simple idea—free for use, modification, resale and fee for services including implementation. There is at least one ERP product (OpenMFG) allowing users to participate in software development, but with software vendor filtering. Open filtering models have not appeared to date.

Among the many open-source ERP vendors, Compiere has most often appeared in many research articles and business reports. Compiere recorded more than 1.2 million downloads of its software and has more than 100 partners in 25 countries.[25] They don't sell software but sell services—security and support. They do not allow just anyone to contribute code—the majority of code contributors are trained partners who understand company's business model. This belongs to limited production/limited filtering model of Web 2.0.

OpenMFG allows community members including customers and partners to get the source code and extend and enhance it. The company, then, bring the enhancements into the product.[26] This could be an example of typical open Production/limited Filtering model.

ERP is an area with great potential for Web Services implementation. Web Services could work as an integration method for the ERP system. More importantly, ERP modules or functionalities could be componentized as Web services. It is already common for ERP vendors to provide services through internet. Customers, especially small and middle size companies, do not need to own ERP software or to invest in the associated infrastructure. They can simply rent and use ERP services.

Evidence of Convergence

Web Services provides a more open system, which should lead to a better market for all participants in the long run. This open market can lead to innovation in enterprise systems. ERP can be viewed as a self-organizing system, whose components can be assembled in an open market accessed over the web.

The composition of Web Services is an important strategic decision in the new software environment. The concept of bundling was demonstrated qualitatively through the ERP market. Vendors can improve their total profit if they open their systems to an open market for non-core components. Their clients can benefit by accessibility to the appropriate functionality of software best fitting their needs (and the right cost).

SOA is a major element of new products by vendors such as SAP, who have been pushing NetWeaver, which promises to open SAP to easier integration and data exchange, making its replacement (MySAP.com) to the core R/3 system more functional.[27] Oracle is doing similar things with its Project Fusion. These also make it more expensive to upgrade, but EIS vendors are forcing some switching through maintenance support policies.[28]

Firms such as Compiere and Tiny ERP provide innovative access to enterprise systems through open source software. Instead of charging millions (many millions for large firms) like SAP and Oracle, open source software is free! One might wonder about their business model, but such vendors seem to be doing well providing service to users. This is massive innovation in the largest software market in the world. While traditional vendors charge more and more and upgrade their software (discontinuing service for old versions) to maximize their revenue streams, strategic innovation has hit even the ERP market through the same open source venue that brought us Linux.

In fact, Linux is being adopted by IBM, which claims that it can save US$400 million annually in programming. This also is an example of innovative use of information technology for strategic innovation.[29]

References

1 B.W. Dearstyne. "Blogs, mashups, & wikis oh my!" *The Information Management Journal* 41, No. 4, 2007, 24–33.

2 S.M. Lee and S. Trimi. "Organizational blogs: Overview and research agenda," *International Journal of Information Technology and Management* 7, No. 2, 2008, 113–19.

3 M. Casteluccio. "A new year, a new Internet," *Strategic Finance* 89, No. 7, 2008, 59–60.

4 D. Tapscott. *Wikinomics: How Mass Collaboration Changes Everything*. New York: Portfolio, 2006.

5 A. Kumnick. "Services oriented architecture: A phoenix rises out of COBOL's ashes," *Canadian Underwriter* 73, No. 2, 2006, 24–6.

6 M.R. Brenner and M.R. Unmehopa. "Service-oriented architecture and Web services penetration in next-generation networks," *Bell Labs Technical Journal* 12, No. 2, 2007, 147–60.

7 Technologies to make Web Services work include XML (Extensible Markup Language), SOAP (Simple Object Access Protocol), WSDL (Web Service Definition Language), UDDI (Universal Description, Discovery and Integration).

8 P. Dube, Y. Hayel and L. Wynter. "Yield management for IT resources on demand: Analysis and validation of a new paradigm for managing computing centres," *Journal of Revenue and Pricing Management* 4, No. 1, 2004, 29–38.

9 L. Bielski. "2006: Building the rational company, *ABA Banking Journal* 98, No. 1, 2006, 35–7.

10 L. Bielski. "Dawn of SOBA?" *ABA Banking Journal* 99, No. 7, 2007, 35–8.

11 E. Schell. "Who's afraid of SOA?" *Multichannel Merchant* 3, No. 3, 2007, 52–3.

12 L. Bielski. *ibid.*

13 J.Y. Sayah and L.J. Zhang. "On-demand business collaboration enablement with web services," *Decision Support Systems* 40, 2005, 107–27.

14 H.K. Bhargava and S. Sundaresan. "Computing as utility: Managing availability, commitment, and pricing through contingent bid auctions," *Journal of Management Information Systems* 21, No. 2, 2004, 201–27.

15 F. Leymann, D. Roller and M.-T. Schmidt. "Web services and business process management," *IBM Systems Journal* 41, No. 2, 2002, 198–210.

16 S. Jones. "Toward an acceptable definition of service," *IEEE Software* 22, No. 3, 2005, 87–93.

17 IBM. "Operating environment essentials for an on demand breakthrough," *E-Business On Demand* March, 2004.

18 N. Serrano and J.M. Sarriegi. "Open source software ERPs: A new alternative for an old need," *IEEE Software* May/June 2006, 94–97.

19 N.G. Carr. "IT doesn't matter," *Harvard Business Review* 81, No. 5, 2003, 41–9.

20 D.L. Olson. *Managerial Issues in Enterprise Resource Planning Systems*. Boston: McGraw-Hill/Irwin, 2004.

21 J. E. Scott and L. Kaindle. "Enhancing functionality in an enterprise software package," *Information & Management* 37, No. 2, 2000, 111–22; C. Dean, "ERP Best Practices Checklist," www.deansystem.com, (2002).

22 D.L. Olson (2004) *ibid*; S. Trimi, S.M. Lee, D.L. Olson and J. Erickson. "Alternative means to implement ERP: Internal and ASP," *Industrial Management and Data Systems* 105, No. 2, 2005, 184–92.

23 M. Santosus. "Impex embraces open-source ERP," *eWeek*, March 26, 2006.

24 N. Serrano and J.M. Sarriegi. 2006, *ibid*.

25 R. Ferguson. "Open-source enterprise push," *eWeek*, 7 January 2008.

26 R. Ferguson. "Open-source ERP grows up," *eWeek*, 10 July 2006.

27 D. Callahan. "SAP moves offerings," *eWeek* 22, No. 21, May 23, 2005.

28 R.B. Ferguson. "ERP friction leaves users cold," *eWeek* 22, No. 4, January 24, 2005.

29 "Something new under the sun: A special report on innovation," *The Economist* 13 October 2007, p. 15.

10 Strategic Innovation

The most important aspect of convergenomics is to apply new developments in convergence to business opportunities in the form of strategic innovation. However, while there are many ideas generated to implement new technologies in new applications, it is a challenge to successfully bring such ideas to profitable fruition. Strategic innovation is fundamental reconceptualization of what a business is for, leading to dramatically different ways to conduct business.[1] This leads to achieving value improvements for customers which in turn leads to high growth for companies that apply strategic innovation.[2] Strategic innovation leading to value innovation makes competition irrelevant by offering fundamentally new and superior buyer value in existing markets, and also creating new markets.[3]

Types of Innovation

Innovation can be viewed in a number of forms.[4] One type focuses on creative ideas to improve products or services. This form of innovation is to build a better mousetrap. While it leads to better things for consumers, it does not alter business models. A second type is to improve processes of the value chain. This form of innovation leads to better efficiency through increased productivity. While it also provides value to consumers, such as lower price and better product quality, it also does not lead to altered business models. A third type is continuous improvement, incremental process improvement typical of programs such as 6-sigma which can be of value to business, but do not lead to the radical change in business models we are interested in.

The last form of innovation is strategic, involving unproven business models. Innovation of this type has been implemented by companies in such different fields as Southwest Airlines (which abandoned the hub-and-spoke business model dominant in major airlines, and competed on direct connections, providing commuters in major routes alternatives to all forms of transportation, to include driving or train-commuting) and IKEA, which revolutionized the furniture industry. Breakthrough strategies can be simple innovations. JC Decaux entered the business of providing municipalities with street furniture for benches, bus shelters, and other facilities at competitive prices. They then strategically innovated by developing complementary income streams through advertising posted on the street furniture.[5]

Table 10.1 Innovation types

Level	Type	Impact
1	Reinventing customer value	Better products and services, speed, customization, emotional gratification (aesthetics, memory, inner satisfaction)
2	Process innovation	Reduced value chain cost and improve quality simultaneously
3	Continuous improvement	E-customers, global customers, customer communities, non-customers
4	Strategic innovation	New ways of doing old tasks, in a much improved manner

Examples of Strategic Innovation

The convergence of technology with business opportunity has been demonstrated in the air travel industry in many forms. Low-cost airlines have reshaped value chains by reinterpreting criteria to measure customer value. Fleet standardization is a means to reduce maintenance costs. Use of cabin crews to clean planes between flights also lowers operating costs. Use of small, low-tax airports enables price advantages.[6] Southwest Airlines and Jet Blue offer a limited menu of services, focusing on providing value in point-to-point business models.[7] Airline innovations can also take advantage of technology development. In Singapore, joint efforts of Singapore Airlines and the Civil Aviation Authority of Singapore have led to the use of biometrics to improve customer flight experience through fully automated seamless travel.[8] Biometrics are used for airline check-in, pre-immigration security checks, and immigration clearance. The clear objective is not simply security, but integrating processes to drive service excellence. This strategic innovation leads to simplified procedures, reduced error, efficient use of self-service technology, greater security, and greater customer service and convenience.

Strategic innovations are also found in banking. Banks serve a mature market, with commodity products and services, and intense competition from many sources, including automobile companies for automobile loans and advisory firms and insurance companies for financial advice, in a highly competitive industry. Product/service innovation options include evolutionary improvement of established business (such as use of web technology for internet banking), process innovations to increase productivity and reduced staffing (such as use of ATMs), value engineering through operational excellence through enterprise system technology, and value innovation. Value innovation creates differentiation from competition through providing superior customer value, leading to new demand.[9] Customer relationship management is a means to seek improved customer value.

Healthcare is a growing sector of the global economy, with many opportunities for strategic innovation. Some hospitals are strategically innovating through the many developments in IT. Examples include web-enabling patient televisions, which can be accessed for medical care by allowing doctors to access patient records over the web rather than needing to bring in a computer. Patients can also utilize web television to complete surveys, order meals, and make purchases.[10] Another example of value innovation in the healthcare industry is medical tourism. Several well-known hospitals, especially Bangkok International Hospital Group and Bumrungrad Hospital in Thailand, have converged their excellent medical services with the world-class hospitality industry, cultural heritage, and Thai government's tourism strategy. The end result is a complete package of patient services from travel plans (visa, airline reservation, airport pickup, and so on.), hospital care, recovery stay in a resort hotel, insurance processing, personal well-being service (daily diet, exercise, massage, and so on.), regular checkups after discharge, and so on.

Strategic Innovation Case[11]

TVS Electronics is an Indian manufacturer of printers, computer keyboards, and related systems. They made a strategic decision to enter the retail automation business with the goal of empowering Indian retailers through retail automation. This initial product-focused innovation was to develop a low cost point-of-sale terminal This system allowed retailers to gain the following IT-enabled functionalities:

- stock management with alarm notification for items low in stock;
- identification of slow-moving items;
- payment handling via cash or credit card;
- sales return handling;
- bill invoicing in multiple languages, customized by locality;
- the ability to change prices;
- hand-held bar code reader;
- web connection.

The product was developed and tested in ten grocery stores in a large city in South India. However, it didn't sell well. The firm then moved to the next level of strategic innovation, focusing on enhancing value for customers:

- Transaction data from ten independent retailers was gathered and analyzed to develop cost-benefit analyses to show to potential customers.
- A simple customer relationship management machine was added to each TVS terminal to conduct market

basket analysis for the most important customers of each retailer.

TVS was able to demonstrate that retailers could add between 2 and 5 per cent to profits through improved inventory management, and tighter control over slow moving items. This improved sales, but not to a level desired by TVS. They therefore moved to the next level of strategic innovation, looking for other revenue sources to improve terminal value to customers.

- loyalty cards were issued to key customers to gather data on purchase patterns;
- a data mining center was created to sell reports to subscribers;
- a wholesale distribution company was created to aggregate demand and enable common sourcing in purchasing;
- an e-shop with call center was created to enable customers to direct-order groceries.

A partner was found to provide a fiber optic backbone to connect all retailers. An IT company was partnered with to provide database management and network management. A research/analytics firm was included in the consortium to data mine and sell reports.

While TVS ended up scaling back this plan initially, it is available to provide IT support to innovative strategic development of their product.

Conclusion

We have stressed that there have been many developments in technology, especially in IT. These advances include enterprise systems and supply chain systems that have widely been used to provide process gains. This in itself is strategically beneficial, but at a level requiring little if any innovation. But these technologies can also be used strategically. The case of TVS was meant to demonstrate that different levels of innovation can be tapped by organizations. All organizations can utilize IT for efficiency gains. However, these gains can be matched by competitors because they can simply buy products from the same vendors.

However, there are opportunities to apply information technologies creatively. OnStar was originally introduced into General Motors vehicles as a safety and emergency service device. Customer requests motivated General Motors to expand functionality to provide the equivalent of on-demand Yellow Pages, providing drivers with information on the cuisine, location, and price lists of restaurants, as well as the ability to unlock car doors when keys were inadvertently left inside.[12] The technological ability to radio signals to unlock cars converged with this need for drivers to recover their vehicles. Wal-Mart has been proactive in the use of RFID technology to improve their control over their inventories and shipments.[13] This is another convergence of radio signal technology, making possible new and more complete ways of tracking inventory. The experiences of Wal-Mart and its vendors in working through the application of new technology are interesting to many. It should provide them yet another innovative way to improve their ability to provide goods at low costs. The TVS case demonstrated the opportunities to expand the use of IT, by focusing on the customer value chain. They also rely on the convergence, in this case of data analysis with customer value improvement. Business success in the future is going to rely upon successfully bringing innovation to the global business market. Our contention is that IT provides a means to obtain strategic innovation. Technology is advancing rapidly on a daily basis. The opportunities are constrained only by human imagination.

References

1 C. Markides. *All the Right Moves: A Guide to Crafting Breakthrough Strategy*. Cambridge, MA: Harvard Business School Press, 1999.

2 B. Schlegelmich, A. Diamantopoulos and P. Kreuz. "Strategic innovation: The construct, its drivers and its strategic outcomes," *Journal of Strategic Marketing* 11, 2003, 117.

3 W.C. Kim and R. Mauborgne. *Blue Ocean Strategy: How to Create Uncontested Market Space and Make the Competition Irrelevant*. Cambridge, MA: Harvard Business School Press, 2005.

4 V. Govindarajan and C. Trimble. "Strategic innovation," *Leadership Excellence* 24, No. 5, 2007, 9–10.

5 B. Moingeon and L. Lehmann-Ortega. "Strategic innovation: How to grow in mature markets," *European Business Forum* 24, 2006, 50–54.

6 *Ibid.*

7 T. Thamara. "Strategic banking innovation," *ABA Bank Marketing* 39, No. 4, 2007, 4–5.

8 L. Heracleous and J. Wirtz, "Biometrics: The next frontier in service excellence, productivity and security in the service sector," *Managing Service Quality* 16, No. 1, 2006, 12–22.

9 T. Thamara (2007), *ibid.*

10 D.J. Sternberg. "Strategic innovation: Some of healthcare's latest innovative technologies might be for you," *Marketing Health Services* 25, No. 2, 2005, 43–4.

11 M.J. Xavier, R. Krishnan and N. Borin. "An integrated model of collaborative value creation for strategic innovation: The case of retail automation in India," *IIMB Management Review* 17, No. 2, 2005, 29–39.

12 "Something new under the sun: A special report on innovation," *The Economist* 13 October 2007, p. 15.

13 C. Fishman. *The Wal-Mart Effect: How the World's Most Powerful Company Really Works – and How It's Transforming the American Economy*. New York: Penguin Books, 2006.

11 Convergence Evolution

Convergences

Convergence is a consequence of the ingenuity and adaptability of the human being. Humans have the ability to learn and creatively transmit the knowledge gained for various practical uses. As we discussed in Chapter 3, convergence can be as simple as combining an eraser with a pencil to being as complicated as combining biology with information technology: biotechnology. Today, we are living through a new wave of convergence that is changing the way societies, organizations, and people work. Convergence revolution is opening new opportunities to create new values and in different ways than was the case with traditional business, the new age of convergenomics. We shall briefly review the evolution process of convergence, which we discussed in detail in Chapter 3, by providing different examples.

Component/Product Convergence

Convergence of components and products has existed throughout time. Clock radios are a result of converging two mature products as components of a new product. Cell phones represent convergence of functions such as internet access, music, photograph, and even telecommunications. The electro motion toothbrush is another good example of convergence obtained by combining electronic skills of the

Braun Oral-B toothbrush and battery technology. Nike and Apple have recently combined the iPod® with gym equipment, allowing health club members to make exercise more entertaining. These two organizations also provided wireless systems embedded in Nike running shoes to communicate with the iPod® Nano for monitoring the heart beat, running speed, and distance the runner is recording while listening to music. LCD technology has allowed development of more realistic video games, better quality movies on airplane flights, and enhanced ability to learn and teach. E-book devices may replace books and newspapers through the use of ambient light from the surrounding area, yielding sharp images readable even in bright sunlight. Once a page is displayed, no further power is needed to keep it visible, as power is only consumed during updating. Newer devices have also been found to be easier on human eyes. This technology has been brought to market by Sony (marketing the Reader), Amazon (Kindle), and Motorola (Motofone).

Functional Convergence

In the 1990s BPR was an innovative approach to cut across functions, developing new work processes, with the intent of identifying the best way of doing things. Many functional organizations, such as R&D, manufacturing, marketing, finance, and human resources, had become isolated silos with limited interfaces. Process innovation through BPR represents functional convergence to develop efficient and lean value chains. Today, enterprise systems are an extreme form of functional convergence through unifying of organizational databases and reporting systems.

Dell Computer streamlined its value chain by disintermediating wholesalers and retailers to provide speedy delivery of PCs to customers. Physicians Mutual shifted from a centralized, mainframe-computing environment to a distributed server-based framework to allow broader access to their information system. As part of the change, business processes were redesigned and a new service-oriented architecture was installed. The system enabled the information technology group at Physician's Mutual to improve service in a rapidly growing environment that was increasingly distributed.[1]

McDonalds was instrumental in making the fast food industry a major component of the US (and even the world's) economy. Prior to 1950s, you could get food in one of many independent restaurants, each with its own décor and cooking style. McDonalds came up with a business functional style, bringing mass production to the food industry. Processes were created to make a limited set of products available in short periods of time, by low-cost labor that followed carefully laid out instructions. This idea led to a massive revolution in food distribution in the US. It had side effects in that it made food obtained during travel less risky from botulism. Advances in ICT, e-business, and global supply chains require streamlined and agile management.

Organizational Convergence

In the era of e-global and digital economy, organizations have the opportunity to collaborate with other firms to enrich their value chain. Strategic alliances and joint ventures with suppliers, component assemblers, distributors, and even competitors are good examples of organizational convergence. Vendor participants in supply chains such as Wal-Mart, Nike and Dell are required to have strong collaborative functions. Interorganizational relationships (IORs) or business-webs (b-webs), where each partner organization brings its own core competence to the value chain, are new business models that agile organizations are using to become and remain competitive.[2] Supply chains are being created that link previously independent organizations, at least temporarily, into cooperative assemblies and draw upon the specialties of participants from around the globe. For example, Nike is simply a brand management firm, while all of its products are produced by its partner organizations all over the world.

Technology Convergence

Since the 1980s, there have been numerous revolutionary convergences of seemly heterogeneous technologies to create new products, services, new processes, or even new technologies.[3] The most prevalent convergences have involved information and communication technology (ICT), biotechnology, engineering, nanotechnology, and artificial intelligence. We have seen liquid-crystal displays (LCDs) replace clock dials and all kinds of meters in automobiles. Science has miniaturized LCDs, enabling convergence with computer devices, leading to brilliant small displays for texts and graphics. The electroluminescence of organic compounds can be utilized to build organic light-emitting diodes (OLEDs), thinner and brighter than LCDs, with wider viewing angles. These devices have thus far appeared as secondary displays on the exterior of mobile phones or in music displays.

In the pharmaceutical industry, many new drugs have been introduced through convergence of biotechnology, agriculture, and chemical engineering. There is the potential to treat patients at the genetic structure level. Neural technology is the convergence of biology with computer science.[4] Neuroscientists are interested in the neocortex as it promises better understanding of how to treat psychiatric disorders and brain diseases, and explain how learning and memory work.

Technology convergence also includes application of modeling approaches taken from engineering and applied to biology. These efforts can lead to biological computers. Convergence of engineering and biotechnology has also yielded remarkable new products and services such as Magnetic Resonance Imaging (MRI) and Positron Emission Tomography (PET) scanning. Further development may lead to atomic force microscopy, enabling scanning at the cellular and sub-cellular level.

Industry Convergence

Industry convergence occurs when two or more previously distinct industries become direct competitors or cooperators or create a new one. Apple's iTunes, which competes with music distributors, is the convergence of IT and music industries. When Walt Disney went into the travel convention business it represented a convergence of entertainment, hotel, and golf industries.

There are three types of industry convergence. Functional convergence is when products from different industries perform the same function(s). Personal computers and television industries both provide some of the same functions, and are becoming more and more substitutes of each other. Complementary convergence occurs when multiple objects of value from different industries are obtained in one location, such as getting travel and lodging services in the same location (as when travel agents sell both airline tickets and hotel rooms). Institutional convergence is when firms behave as though products of two industries are linked. While complementary convergence focuses on the product being developed, institutional convergence focuses on the organizational structure. The banking industry has converged with the mortgage industry, so that they can provide both checking deposit services as well as mortgage loans. The edutainment industry is the result of convergence between education and entertainment. Some oil industry firms claim that they are in the energy business, allowing them the latitude to develop petroleum as well as hybrid fuels. Digital convergence of IT and broadcasting is another example of institutional convergence. ENPS shifted its focus from output at given intervals (such as the evening news) to pools of content distributed in multiple ways. Editorial work is unified to serve all media, providing output for television, radio, print, and the Web.

Open-source Convergence

The advent of Web 2.0 opened another level of convergence, which was not discussed in Chapter 3. As discussed in Chapter 9, Web 2.0 includes three key properties: openness, participation, and collaboration.[5] Openness in the Web 2.0 era, in addition to open-source, includes open innovation, knowledge sharing, crowd-sourcing, and so on. Open innovation is a new convergence phenomenon that organizations need to be successful through relying on interactive participation by multiple users to generate new and better solutions. Crowdsourcing is inviting the public to take on the tasks that were traditionally performed by employees or contractors. The perceived benefits of crowd-sourcing (Wikipedia) are a comparatively low cost to explore given problems, payment by results, and the ability to tap a wide range of talent.

In Web 2.0, information consumers are becoming information providers (user generated content). Highly skilled hobbyists (prosumers) before Web 2.0, are being replaced by Pro–Ams who are creating contents of value that contribute to society and the economy. The primary motivation for Pro–Ams to participate in open innovation is attention. How to create new business models based on the convergence of attention motivated open collaboration is indeed critical.

Biological and Artificial Systems Convergence

The convergence of biological and artificial systems represents the highest level of convergence. As such, it represents the cutting edge of development, where ideas from the prior levels of convergence can be utilized to lead radically new methods and products. Ubiquitous life (u-life) involves a state where our lives are supported and even managed by a combination of technologies and systems. These advances are changing the life style of people through miniaturization, visualization, manipulation, and evaluation.

Miniaturization leads to application in a number of areas. We are all familiar with the advances in computer chip technology, and Moore's Law, characterizing the geometric rate of improvement in computational power in an ever-decreasing size of the chip. Miniaturization also has impacts in health care, through gene splicing and molecular biology.

As reviewed in Chapter 7, visualization advances come from molecular technology as well. X-ray crystallography and atomic force microscopes have evolved, giving scientists the ability to peer into the extreme nanospace of nature. There are also commercial applications, in the pharmaceutical industry, the petroleum industry, and other useful applications. These developments offer promise of convergence of bio/artificial systems.

Manipulation extends the knowledge gained through visualization. Thus biological and artificial systems convergence leads to better abilities to train those dealing with critical activities to learn through virtual reality and tactile feedback. This can be extremely useful in fields such as surgery on the nervous system, as well as in critical operations such as nuclear plant operation.

Evaluation is gaining value from technology, as with the genome project. As mentioned in Chapter 7, other technologies include microarrays, biotechnology hybrids, and information technology. This has enabled micro-quality control, and the ability to better predict when specific parts in mechanical products such as automobiles or refrigerators are wearing out. This again can offer great opportunities in critical fields such as medicine, nuclear power generation, or airline transportation. It also could have beneficial effect in fields such as detection of terrorist devices. And beneficial products can be developed for less critical domains as well, giving consumers the opportunity for new products they find of value.

Gains from the convergence of engineering and biological studies were found at the University of California, Santa Barbara with the aid of Microsoft Corporation,[6] leading to new technologies to enable people to interact with computers in more powerful ways. This is a direct application of virtual reality, which has been around for decades, but new convergences of biological and artificial systems make it possible to harness more of virtual reality's potential.

Another intriguing area of the potential for biological and artificial system convergence is the integration of Western and Chinese medicine. These medical approaches developed in isolation for millennia. But there is greater realization that the two radically different fields of expertise can learn from each other, and lead to greater opportunities for better medical care and wellness.

Table 11.1 recapitulates some of the convergence examples discussed previously.

Table 11.1 Examples of convergence outcomes

Level	Manufacturing	Service
Product convergence	Cell phones (telephone, information access, camera, games) Coffee (soy milk, decaffeinated, diet, frappacinos)	Jazzercise
Functional convergence	Lean manufacturing Enterprise resource planning systems	McDonalds' hamburgers (fast service, consistent quality, entertainment for kids)
Organizational convergence	Steel (movement from mass production to minimills)	Banks (from checking accounts to include savings, mortgages, bill paying, many other services)
Technology convergence	Hybrid cars	Cell phones (communication, music, e-mail, photos, and so on.)
Industry convergence	Starbucks Political ads (from stump speeches and posters to a transformation of the television industry)	e-Bay (transformed want-ads and flea markets to get rid of white elephants to an e-commerce revolution)
Open-source convergence	Software systems to support manufacturing (Linux, open-source ERP systems)	Web services linked in service oriented architecture
Bio/artificial systems convergence	Medical testing (X-rays to MRIs and beyond)	Chinese medicine

Strategic Innovation in Convergenomics

The convergence revolution is opening new opportunities to create value in different ways than the traditional business thinking. The traditional management approaches emphasize reinforcing or optimizing the existing core competence of the firm as a stand-alone organization. Examples of these approaches include benchmarking, continuous improvement, differentiation, focus, cost leadership, global strategy, and customer-centric management. The traditional management strategy involves improving efficiency in the value chain by emphasizing economies of scale, economies of scope, and market expansion. While these practices are good when executed properly to increase the firm's share in a zero-sum game, they would not allow the firm to reinvent itself to explore blue oceans.[7]

In the new era of convergenomics, organizations are not an isolated and independent entity but a member of the wide open global community for collaboration or partnerships. In convergenomics, organizations pursue not only economies of scale and scope but also economies of expertise and collaboration.[8]

Strategic innovation is fundamental reconceptualization of how an organization creates value in dramatically different ways. Innovation can be classified in a number of ways. In Table 11.2, innovation is classified in terms of its level and purpose in three levels where organizations pursue innovation for value creation.

The first level involves improving the efficiency of the value chain by dramatically redesigning the end-to-end value chain architecture. In this phase, innovation involves application of many proven or new approaches for procurement, transformation, and distribution channel innovation as follows:

- procurement innovation: just-in-time system, outsourcing, electronic markets;
- transformation innovation: lean manufacturing, production versus assembly, outsourcing, bundling, debundling, and rebundling;
- distribution channel innovation: disintermediation, outsourcing, direct marketing via the demand chain.

Table 11.2 Innovation levels

Level	**Purpose**	**Value Creation Objective**
1	Process innovation	Improve the value chain architecture Procurement Transformation Distribution
2	Customer value innovation	Enrich and reinvent customer value Cost Quality Speed Customization Emotional needs
3	Customer base innovation	Redefine and create the customer base End user E-customer Customer communities Global customers Non-customers

The second level of innovation involves customer value creation. The traditional values of cost and quality are simply market entry conditions. They are no longer sufficient or sustaining customer values. Today, organizations must pursue new convergence to provide speed (responding almost instantaneously to customer needs), customization (providing exactly or beyond customers' expectations of what they want), and satisfy customers' emotional needs (beauty, elegance, memory, hope, sense of security, and so on).

The third level of innovation deals with redefining the customer base. The traditional definition of the customer is simply the end user of a product or service. However, in the e-global economy, customers can be those who do not actually consume the product or who are physically in contact with the product. E-customers, global customers, customer communities, and current non-customers (who may become customers in the future) should all be included in the customer base. Through convergence of technologies and new methods, organizations must find ways to create demand or develop new markets in the blue ocean. Figure 11.1 demonstrates the relationships of these concepts.

Innovation can be any combination of exploitation of what is known and exploration of what is unknown. Thus, we can classify strategic innovation also based on the type of approach used (Figure 11.2). The first type can be classified as *evolutionary* or optimizing innovation where continuous improvement of the existing core competencies is obtained (Kaizen is the main thrust—an evolutionary development). On the other extreme (of classification) is *revolutionary* innovation, where new frontiers are explored to create new core competencies. Two other types of innovation come as a combination between the two approaches. The ultimate innovation type is where both exploitation and exploration are high. We call this ubiquitous innovation based on convergenomics. The optimum balance between exploitation and exploration for any given organization depends on several factors such as the industry type; size, age, management vision, product lines, and DNA of the organization.

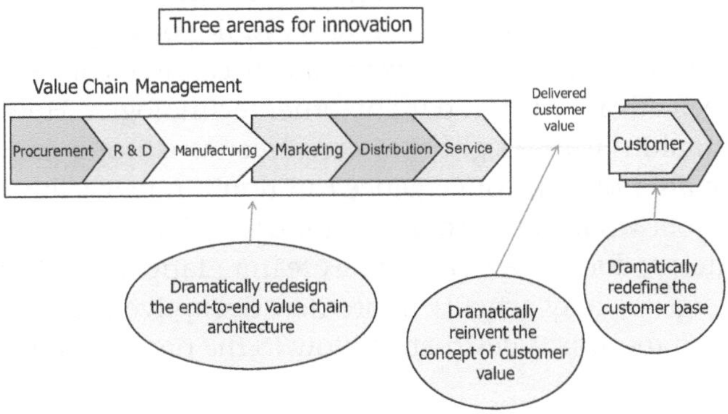

Figure 11.1 The value chain and innovation levels. Adapted from Gupta and Govindarajan (2004)

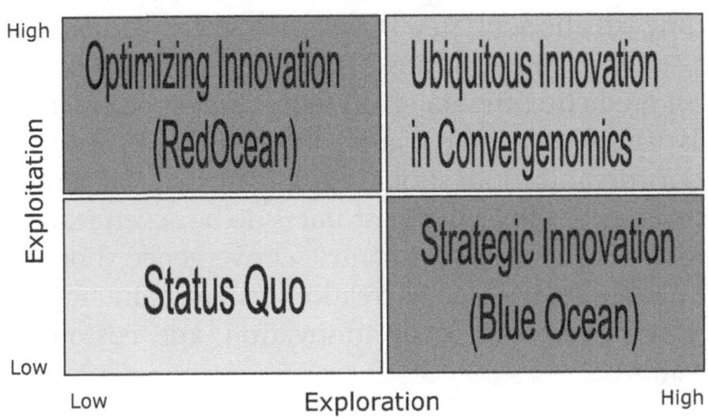

Figure 11.2 Strategic innovation classification near

Success Factors in Convergenomics

We live in the exciting and dynamic age of convergenomics. Many new values can be created to develop competitive advantage for the organization. Also, new products, services, markets, opportunities for collaboration, knowledge sharing, and even creating the feelings of joy and happiness can become realities because of successful convergence. The ubiquitous life (u-life) style is feasible because of convergence between advanced IT and many other technologies. However, we should not lose sight that convergence is not a panacea. As a matter of fact, convergence can result in new values that overshoot what the customers want or need, thus resulting in counterproductive effects such as feature fatigue. Such failed convergence may actually invite new disruptive technologies that can result in a destructive blow to the organization.

To be successful in convergenomics and sustain competitive advantage, organizations must explore convergence opportunities with a firm belief that convergence is basically a "consumer science" rather than "technology engineering." In other words, convergence should be customer centric, pursuing creation of new values for the customer. As we discussed in Chapter 3, organizations need to be constantly alert in predicting the trajectory of convergence based on the trends of customer habits, values, tastes, life style, and so on. In addition, organizations should explore new convergences that can create new values that customers do not even realize they are missing. The customer centric convergence should focus on the three areas of value creation: value chain architecture innovation, customer value innovation, and customer base innovation.

Value chain architecture innovation involves new approaches to make the chain leaner, more efficient, and more cost-effective. New purchasing ideas, sourcing innovation, streamlining the distribution channel (for example, deintermediation), and so on. through convergence are good examples. The customer centric approaches to new customer value creation and customer base expansion involve convergence through bundling-debundling-rebundling, as well as exploratory approaches, for product/service upgrades, product/service life cycle extension, new products/services, new customer values, and new markets. The following are good examples of successful convergences.

1. Product and product convergence: providing additional functions, convenience, reduced cost, space saver, enjoyment, comfort and feeling of security, and so on, as new customer values.

    - cell phone with camera
    - navigation system in the automobile
    - washer/dryer combination.

2. Product and service convergence: bundling a product with services.

    - General Motors automobiles—GM Telematics
    - Apple iPod®—music downloading service for iPod owners.

3. Service and service convergence: creating new customer values by bundling services.

- KT Bismeka—Korea Telecom provides ERP functionality rental service to its telephone customers
- Disney World—A packaged service for amusement park, hotel, resort, golf, and other services
- U-community—housing facility, entertainment and recreation centers, shopping, education, restaurants, and so on.

Another key to success in convergenomics, in addition to predicting the trajectory of trends, is becoming the first mover or the blue ocean pioneer. Convergence is typically not a binary event of yes or no. Rather it is the degree of convergence, balancing the bundling of technologies/products/services, or harmonizing value creators. The four dimensions of blue ocean strategy of eliminate, reduce, raise and create can be applied to convergence: eliminate those the industry considers to not add much value to the customer, decrease those that add limited values, raise those the customer values, and create those that will provide new values to the customer and expand the customer base.

Conclusion

The world is becoming flatter because of the combined effects of new developments in geopolitics, advances in technologies, and strategic innovation approaches. Nevertheless, this road has been anything but smooth. There are widening economic and digital divides. While the broadband internet is becoming more widely available throughout the world, there are millions of people who are trying to survive on less than US$1 a day. The new age of convergenomics has brought new value creating opportunities and business models. Strategic innovation is a requirement for firms to survive in convergenomics.

The era of convergenomics has opened a new blue ocean for developing new products/services, design lean value chains, create new customer values, and expand the customer base. Exploring and predicting the trajectory of new convergences are an important part of strategic innovation for any organization. The first mover advantage in the deep blue sphere of convergenomics will yield enormous wealth creating opportunities to ubiquitous innovators. In the world of convergenomics, opportunities for strategic innovation are constrained only by human imagination and creativity.

References

1 Anonymous. "Storage insurance," *Health Management Technology* 27, No. 10, 58, 60, 66, 2006.

2 D. Tapscott and A.D. Williams. *Wikinomics: How Mass Collaboration Changes Everything*. New York: Portfolio Wikinomics, 2008.

3 Bacon, D. and Leung, D. "Toward a world with quantum computers," *Communications of the ACM* 50:9, 2007, 55–59.

4 Tablot, D. "IBM: The computer brain," *Technology Review*, September 2005, 52.

5 Tapscott, D. *Wikinomics: How Mass Collaboration Changes Everything*. New York: Portfolio, 2006.

6 University of California – Santa Barbara. "Is that tea real?" *Convergence: The Magazine of Engineering the Sciences at UC Santa Barbara* 7: Spring, 2007, 3–5.

7 W.C. Kim and R. Mauborgne. *Blue Ocean Strategy: How to Create Uncontested Market Space and Make the Competition Irrelevant.* Cambridge, MA: Harvard Business School Press, 2004.

8 A.K. Gupta and V. Govindarajan. *Global Strategy and Organization*. Hoboken, NJ: Wiley & Son, 2004.

Bibliography

Books

Allison, G. *Nuclear Terrorism: The Ultimate Preventable Catastrophe*. New York: Henry Holt and Company, 2004.

Asthana, R., Kumar, A. and Dahotre, N.B. *Materials Processing and Manufacturing Science*. Boston: Elsevier Academic Press, 2006.

Brown, L.R. *Eco-Economy: Building an Economy for the Earth*. New York: W. W. Norton & Company, Inc., 2001.

Burke, J. *Connections*. New York: Simon and Schuster, 2007.

Castells, M. *End of Millenium*, 2nd ed. Malden, MA: Blackwell Publishing, 2000.

Castells, M. *The Rise of the Network Society*, 2nd ed. Malden, MA: Blackwell Publishing, 2000.

Christensen, C.M. *The Innovator's Dilemma: When New Technologies Cause Great Firms to Fail*. Cambridge, MA: Harvard Business School Press, 1997.

Dosi, G., Freeman, C., Nelson, R., Soete, L. and Silverberg, G. *Technical Change and Economic Theory*. London: Pinter, 1988.

Drexler, K.E. *Nanosystems: Molecular Machinery, Manufacturing, and Computation*. Chichester: John Wiley & Sons, 1992.

Drucker, P. *Management Challenges for the 21st Century*. New York: Collins, 2001.

Drucker, P. *Managing in the Next Society*. New York: St. Martin's Press, 2002.

Duncan, D.E. *Masterminds: Genius, DNA, and the Quest to Rewrite Life*. New York: Harper Perennial, 2005.

Fishman, C. *The Wal-Mart Effect: How the World's Most Powerful Company Really Works – and How It's Transforming the American Economy*. New York: Penguin Books, 2006.

Fishman, T.C. *China Inc.: How the Rise of the Next Superpower Challenges America and the World*. New York: Scribner, 2005 (2006).

Friedman, T.L. *The World is Flat: A Brief History of the Twenty-First Century*. New York: Farrar, Strauss and Giroux, 2005.

Fukuyama, F. *The End of History and the Last Man*. New York: The Free Press, 1992.

Fukuyama, F. *Our Posthuman Future: Consequences of the Biotechnology Revolution*. New York: Picador, 2002.

Gasman, L. *Nanotechnology Applications and Markets*. Boston: Artech House, 2006.

Gross, M. *Travels to the Nanoworld: Miniature Machinery in Nature and Technology*. Cambridge MA: Perseus Books, 1999.

Gupta, A.K. and Govindarajan, V. *Global Strategy and Organization*. Hoboken, NJ: Wiley & Son, 2004.

Harrington, H.J., Essing, E.K.C. and van Nimwegen, H. *Business Process Improvement Workbook: Documentation, Analysis, Design, and Management of Business Process Improvement*. New York: McGraw-Hill, 1997.

Hawken, Paul. *The Ecology of Commerce: A Declaration of Sustainability*. New York: Harper Business, 1993.

Holland, J.H. *Adaptation in Natural and Artificial Systems*. Cambridge MA: MIT Press, 1992.

Holland, J.H. *Hidden Order: How Adaptation Builds Complexity*. Cambridge MA: Perseus Books, 1995.

Holzner, S. *How Dell Does It: Using Speed and Innovation to Achieve Extraordinary Results*. New York: McGraw-Hill, 2006.

IBM. The Enterprise of the Future, Global CEO Study, 2008.

Johnson, C. *Blowback: The Costs and Consequences of American Empire*. New York: Owl Books, 2000.

Kauffman, S. *Investigations*. New York: Oxford University Press, 2000.

Kim, W.C. and Mauborgne, R. *Blue Ocean Strategy: How to Create Uncontested Market Space and Make the Competition Irrelevant*. Cambridge, MA: Harvard Business School Press, 2005.

Klare, Michael T. *Resource Wars: The New Landscape of Global Conflict*. New York: Henry Holt and Company, 2001.

Kurzweil. R. *The Age of Spiritual Machines: When Computers Exceed Human Intelligence*. New York: Penguin Books, 2000.

Kurzweil, R. *The Singularity is Near: When Humans Transcend Biology*. New York: Penguin Books, 2006.

Lee, S.M. and Olson, D.L. *Introduction to Management Science*. Cincinnati: Southwest, 2006.

Levitt, S.D. and Dubner, S.J. *Freakonomics*. New York: William Morrow, 2005.

Mandeles, M.D. *The Future of War: Organizations as Weapons*. Washington DC: Potomac Books, 2005.

Markides, C. *All the Right Moves: A Guide to Crafting Breakthrough Strategy*. Cambridge, MA: Harvard Business School Press, 1999.

Meadows, D., Randers, J. and Meadows, D. *Limits to Growth: The 30-Year Update*. White River Junction, VT: Chelsea Green Publishing Company, 2004.

Menzies, G. *1421: The Year China Discovered America*. New York: Harper Perennial, 2004.

Meyer, C. and Davis, S. *It's Alive: The Coming Convergence of Information, Biology, and Business*. New York: Crown Business, 2003.

O'Leary, D.E. *Enterprise Resource Planning Systems: Systems, Life Cycle, Electronic Commerce, and Risk* Cambridge: Cambridge University Press, 2000.

Oliver, R.L. *The Business of Biotech and How to Profit From It*. New York: McGraw-Hill, 2003.

Olson, D.L. *Managerial Issues in Enterprise Resource Planning Systems*. Boston: McGraw-Hill/Irwin, 2004.

Perrow, C. *Normal Accidents: Living with High-Risk Technologies*. Princeton, NJ: Princeton University Press, 1999.

Phillips, M.I. (ed.) *Gene Therapy Methods*. San Diego: Academic Press, 2002.

Porter, M. *Competitive Advantage: Creating and Sustaining Superior Performance*. New York: Free Press, 1998.

Porter, M.E. and Teisberg, E.O. *Redefining Health Care: Creating Value-Based Competition on Results*. Cambridge, MA: Harvard Business School Press Books, 2006.

Pringle, P. *Food, Inc.: Mendel to Monstanto – The Promises and Perils of the Biotech Harvest*. New York: Simon & Schuster Paperbacks, 2003.

Rapley, R. and Harbron, S. (eds) *Molecular Analysis and Genome Discovery*. Chichester: J. Wiley, 2004.

Roberts, Paul. *The End of Oil: On the Edge of a Perilous New World*. New York: Mariner Books, 2004.

Scheer, H. *The Solar Economy: Renewable Energy for a Sustainable Global Future*. London: Earthscan, 2002, orig. in German 1999.

Schumpeter, J.A. *Capitalism, Socialism and Democracy*. New York: Harper and Row, 1942.

Schumpeter, J.A. *The Theory of Economic Development, 5th ed.* New York: Oxford University Press, 1967.

Tapscott, D. *Wikinomics: How Mass Collaboration Changes Everything*. New York: Portfolio, 2006.

Tapscott, D. and Williams, A.D. *Wikinomics: How Mass Collaboration Changes Everything*. New York: Portfolio Wikinomics, 2008.

Weart, S.R. *The Discovery of Global Warming*. Cambridge, MA: Harvard University Press, 2003.

Articles

Achabal, D.D., McIntyre, S.H., Smith, S.A. and Kalyanam, K. "A decision support system for vendor managed inventory," *Journal of Retailing* 76, No. 4, 2000, 430–54.

Andel, T. "Manage inventory, own information," *Transportation and Distribution* 37, 1996, 55–8.

Anon., "A route to higher profit," *Railway Gazette International* 162, No. 3. March 2006, 130.

Bacon, D. and Leung, D. "Toward a world with quantum computers," *Communications of the ACM* 50:9, 2007, 55–59.

Baggott, E. "A wealth deferred," *Harvard International Review* 28, No. 3, 2006, 28–30.

Barratt, M. "Positioning the role of collaborative planning in grocery supply chains," *The International Journal of Logistics Management* 14, No. 2, 2003, 53–66.

Bhargava, H.K. and Sundaresan, S. "Computing as Utility: Managing Availability, Commitment, and Pricing through Contingent Bid Auctions," *Journal of Management Information Systems* 21, No. 2, 2004, 201–27.

Bhaskaran, K. and Schmidt, M.-T. "WebSphere Business Integration: An architectural overview," *IBM Systems Journal* 43, No. 2, 2004, 238–54.

Bielski, L. "2006: Building the rational company," *ABA Banking Journal* 98, No. 1, 2006, 35–7.

Bielski, L. "Dawn of SOBA?" *ABA Banking Journal* 99, No. 7, 2007, 35–8.

Brännback, M., de Heer, A.J. and Wicklund, P. "The convergence of the pharmaceutical and the food industry through functional food: Strategic change and business opportunity or an illusion?" *Pharmaceuticals Policy and Law* 5 (2002) 63–78.

Brenner, M.R. and Unmehopa, M.R. "Service-oriented architecture and Web services penetration in next-generation networks," *Bell Labs Technical Journal* 12, No. 2, 2007, 147–60.

Brinn, Cy. Improving yield management, *Broker Magazine* 7, No. 5, August 2005, 68–9.

Brown, T.A. and Bukovinsky, D.M. "ECR and grocery retailing: An exploratory financial statement analysis," *Journal of Business Logistics* 22, No. 2, 2001, 77–90.

Callahan, D. "SAP moves offerings," *eWeek* 22, No. 21, May 23, 2005.

Capell, K. "Of neural sound waves and raging hormones," *Business Week* 4028, 4/2/2007, 69.

Carr, N.G. "IT doesn't matter," *Harvard Business Review* 81, No. 5, 2003, 41–9.

Casteluccio, M. "A new year, a new Internet," *Strategic Finance* 89, No. 7, 2008, 59–60.

Cetinkaya, S. and Lee, C.Y. "Stock replenishment and shipment scheduling for vendor-managed inventory systems," *Management Science* 46, No. 2, 2000, 217–32.

Chang, K. "I.B.M. Creates a Tiny Circuit out of Carbon," *New York Times* 27 August 2001.

Chon, B.S., Choi, J.H., Barnett, G.A., Danowski, J.A. and Joo, S.H. "A structural analysis of media convergence: Cross-industry mergers and acquisitions in the information industries," *Journal of Media Economics* 16, No. 3, 2003, 141–57.

Coase, R.H. "The nature of the firm," *Econometrica* 4, 1937, 386–405.

"Convergence makes a comeback," IBE January–February 2006, 36–37.

Convergence: The Magazine of Engineering and the Sciences at UC Santa Barbara, "Is that tea real?" 7: Spring, 2007, 3–5.

Cooke, J.A. "VMI: Very mixed impact?" *Logistics Management and Distribution Report* 37, No. 12, 1998, 51.

Corrado, C., Lengermann, P., Bartelsman, E.J. and Beaulieu, J.J. "Sectoral productivity in the United States: Recent developments and the role of IT," *German Economic Review* 8, No. 2, 2007, 188–210.

Davenport, T.H. "Putting the enterprise into the enterprise system," *Harvard Business Review* (July–August 1998), 121–31.

Dearstyne, B.W. "Blogs, mashups, & wikis oh my!" *The Information Management Journal* 41, No. 4, 2007, 24–33.

Dosh, C. Face scarce space, *Business Travel News* 23, No. 4, 3/6/2006, 23–31.

Dube, P., Hayel, Y. and Wynter, L. "Yield management for IT resources on demand: Analysis and validation of a new paradigm for managing computing centres," *Journal of Revenue and Pricing Management* volume 4, No. 1, 2004, 29-38.

Duysters, G. and Hagedoorn, J. "Technological convergence in the IT industry: The role of strategic technology alliances and technological competencies," *International Journal of the Economics of Business* 5, No. 3, 1998, 355–68.

The Economist. "Displays to keep an eye on," 10 March 2007, pp. 4, 6.

The Economist. "Let's get physical," 10 March 2007, p. 6.

The Economist. "Something new under the sun: A special report on innovation," 13 October 2007, p. 15.

The Engineer. "Life less ordinary," 18 June–1 July, 2007, 30–31.

Ferguson, R. "Open-source enterprise push," *eWeek*, 7 January 2008.

Ferguson, R. "Open-source ERP grows up," *eWeek*, 10 July 2006.

Ferguson, R.B. "ERP friction leaves users cold," *eWeek* 22, No. 4, January 24, 2005.

Fliedner, G. "CPFR: An emerging supply chain tool," *Industrial Management & Data Systems* 103, No. 1, 2003, 14–21.

Frankel, R., Goldsby, T.J. and Whipple, J.M. "Grocery industry collaboration in the wake of ECR," *The International Journal of Logistics Management* 13, No. 1, 2002, 57–72.

Gattorna, J. Supply chains are the business, *Supply Chain Management Review* 10, No. 6, 2006, 42–9.

Ghemawat, P. "Managing Differences: The Central Challenge of Global Strategy," *Harvard Business Review* 85, No. 3, 2007, 59-68.

Gilleo, K. "The new convergence: Bio, nano and MEMS appear a 'magic mix' of science and technology," *Circuits Assembly* 17, No. 2, Feb, 2006, 32.

Govindarajan, V. and Trimble, C. "Strategic innovation," *Leadership Excellence* 24, No. 5, 2007, 9–10.

Grover, V. and Saeed, K. "The telecommunication industry revisited: The changing pattern of partnerships," *Communications of the ACM* 46, No. 7, 2003, 119–25.

Hammer, M. and Stanton, S. "How process enterprises really work," *Harvard Business Review* November–December 1999, 108–18.

Heracleous, L. and Wirtz, J. "Biometrics: The next frontier in service excellence, productivity and security in the service sector," *Managing Service Quality* 16, No. 1, 2006, 12–22.

Holweg, M., Disney, S., Holström, J. and Småros, J. "Supply chain collaboration: Making sense of the strategy continuum," *European Management Journal* 23, No. 2, 2005, 170–81.

IBM. "Operating environment essentials for an on demand breakthrough," *E-Business On Demand* March, 2004.

Ireland, R. and Bruce, R. "CPFR: Only the beginning of collaboration," *Supply Chain Management Review* 4, No. 4, 2000, 80–8.

Johnson, Emma. "Yielding results," *Journal of Property Management* 71, No. 1, Jan/Feb 2006, 43–47.

Jones, S. "Toward an acceptable definition of service," *IEEE Software* 22, No. 3, 2005, 87–93.

Kennedy, J.V. "Nanotechnology: Breaking through the next big frontier of knowledge," *Indiana Business Review* Summer 2007, 7–9.

Kesseldahl, A. "Apple and Nike, running mates," *Business Week Online* 5/9/2006, 9.

Kumnick, A. "Services oriented architecture: A phoenix rises out of COBOL's ashes," *Canadian Underwriter* 73, No. 2, 2006, 24–6.

Lee, H.L., So, K.C. and Tang, C.S. "The value of information sharing in a two-level supply chain," *Management Science* 46, No. 5, 2000, 626–43.

Lee, S.M. and Trimi, S. "Organizational blogs: Overview and research agenda," *International Journal of Information Technology and Management* 7, No. 2, 2008, 113–19.

Leymann, F., Roller, D. and Schmidt, M.-T. "Web Services and Business Process Management," *IBM Systems Journal* 41, No. 2, 2002, 198–210.

Lyytinen, K. and Yoo, Y. "Research commentary: The next wave of nomadic computing," *Information Systems Research* 13, No. 4, 2002, 377–88.

Magin, S., Algesheimer, R., Huber, F. and Herrmann, A. "The impact of brand personality and customer satisfaction on customer's loyalty: Theoretical approach and findings of a causal analytical study in the sector of Internet Service Providers," *Electronic Markets* 13, No. 4, 2003, 294–308.

Malhotra, A. and Gupta, A.K. "An investigation of firms' strategic response to industry convergence," *Academy of Management Proceedings* G1–G6, 2001.

Micheau, V.A. "How Boeing and Alcoa implemented a successful vendor managed inventory program," *The Journal of Business Forecasting* Spring 2005, 17–19.

Minahan, T. "Enterprise resource planning: Strategies not included," *Purchasing* 125, No. 1 (July 16, 1998), 112–27.

Moingeon, B. and Lehmann-Ortega, L. "Strategic innovation: How to grow in mature markets," *European Business Forum* 24, 2006, 50–4.

Nazzaro, R.M. and Hephner, T.S. "Valles Caldera: Trust has made some progress, but needs to do more to meet statutory goals: *GAO-06-98," GAO Reports* 11/16/2005, 1–47.

Oshri, I. and Weeber, C. "Cooperation and competition standards-setting activities in the digitization era: The case of wireless information devices," *Technology Analysis & Strategic Management* 18, No. 2, 2006, 265–83.

Papazoglou, M.P. and Georgakopoulos, D. "Service-oriented computing," *Communications of the ACM* 46, No. 10, October 2003, 24–8.

"Partners in wealth," *The Economist* 378, No. 8461, 1/12/2006, 16–17.

Rettig, C. "The trouble with enterprise software," *MIT Sloan Management Review* Fall, 2007, 21–7.

Santosus, M. "Impex embraces open-source ERP," *eWeek*, March 26, 2006.

Sayah, J.Y. and Zhang, L.J. "On-demand business collaboration enablement with web services," *Decision Support Systems* 40, 2005, 107–27.

Schell, E, "Who's afraid of SOA?" *Multichannel Merchant* 3, No. 3, 2007, 1, 52–3.

Schlegelmich, B., Diamantopoulos, A, and Kreuz, P., "Strategic innovation: The construct, its drivers and its strategic outcomes," *Journal of Strategic Marketing* 11, 2003, 117.

Schultz, D.E. "Subliminal ad notions still resonate today," *Marketing News* 41, No. 5, 3/15/2007, 5, 9.

Schwartz, D. "Mobilizing for mobility: Delivering the 'quadruple play' with fixed mobile convergence," *www.cedmagazine.com*, September 2005, 30–1.

Scott, J.E. and Kaindle, L. "Enhancing functionality in an enterprise software package," *Information & Management* 37, No. 2, 2000, 111–22.

Serrano, N. and Sarriegi, J.M. "Open source software ERPs: A new alternative for an old need," *IEEE Software* May/June 2006, 94–97.

Smith, B.C., Günther, D.P., Rao, B.V. and Ratliff, Richard M. "E-commerce and operations research in airline planning, marketing, and distribution," *Interfaces* 31, No. 2, March–April 2001, 37–55.

Sterman, J.D. "Modeling managerial behavior: Misperceptions of feedback in a dynamic decision making experiment," *Management Science* 35, No. 3, 1989, 321–39.

Sternberg, D.J. "Strategic innovation: Some of healthcare's latest innovative technologies might be for you," *Marketing Health Services* 25, No. 2, 2005, 43–4.

"Storage insurance," *Health Management Technology* 27, No. 10, 58, 60, 66, 2006.

Talbot, D. "IBM: The computer brain," *Technology Review*, September, 2005, 52.

Taniguchi, N. "On the basic concept of 'Nano-Technology'," *Proc. Intl. Conf. Prod. Eng. Tokyo, Part II*, Japan Society of Precision Engineering, 1974.

Taylor, D. "Measurement and analysis of demand amplification across the supply chain," *The International Journal of Logistics Management* 10, No. 2, 1999, 55–70.

Thamara, T. "Strategic banking innovation," *ABA Bank Marketing* 39, No. 4, 2007, 4–5.

Trimi, S., Lee, S.M., Olson, D.L. and Erickson, J. "Alternative means to implement ERP: Internal and ASP," *Industrial Management and Data Systems* 105, 2005, 184–92.

Tynan, D. "Cars that are smarter than you are," www.pcworld.com May 2006, 45.

Van den Berghe L. and Verweire, K, "Convergence in the financial services industry," *The Geneva Papers on Risk and Insurance* 25, No. 2, 2000, 262–72.

Vergin, R.C. "An examination of inventory turnover in the Fortune 500 industrial companies," *Production and Inventory Management Journal* 39, 1998, 51–6.

Weil, M. "Managing to win," *Manufacturing Systems* 17, No. 11, 1999, 14.

Wonglimpiyarat, J. "Convergence for innovation," *Asia-Pacific Tech Monitor* Sep–Oct 2006, 16–24.

Xavier, M.J., Krishnan, R. and Borin, N. "An integrated model of collaborative value creation for strategic innovation: The case of retail automation in India," *IIMB Management Review* 17, No. 2, 2005, 29–39.

Xavier, M.J., Krishnan, R. and Borin, N. "Something new under the sun: A special report on innovation," *The Economist* 13 October 2007, p. 15.

Xu, K. and Dong, Y. "Information gaming in demand collaboration and supply chain performance," *Journal of Business Logistics* 25, No. 1, 2004, 121–44.

Yoffie, D.B. "Competing in the age of digital convergence," *California Management Review* 38, No. 4, 1996, 31–53.

Zhu, J. "Using turndowns to estimate the latent demand in a car rental unconstrained demand forecast," *Journal of Revenue & Pricing Management* 4, No. 4, January 2006, 344–53.

Index

For Product Safety Concerns and Information please contact our EU representative GPSR@taylorandfrancis.com Taylor & Francis Verlag GmbH, Kaufingerstraße 24, 80331 München, Germany